A human being is a part of the whole called by us universe, a part limited in time and space. He experiences himself, his thoughts and feelings as something separated from the rest, a kind of optical delusion of his consciousness.
Albert Einstein

KDP

Copyright © 2016 by Paul Richard Hillson

Cover design: Paul Richard Hillson

Second Revised Edition	January 2018
Third Revised Edition	February 2019
Fourth Edition	Revised June 2019
Fifth Revised Edition	February 2022

Copy Editor / Proofreader - Von Lerma

All rights reserved. No part of this book may be reproduced or transmitted in any form or by any means, electronic or mechanical, including recording, photocopying, or by information storage and retrieval system, without the written permission from the author.

First printing, December 2016
ISBN 9781520154695

THE SUBTLE SHIFT

The Fundamental Mechanism
Paving the Road to the
Extinction *of* Humanity

By

PAUL RICHARD HILLSON

Second Revised Edition   January 2018
Third Revised Edition   February 2019
Fourth Revised Edition   June 2019
Fifth Revised Edition   February 2022

Copyright © 2016

http://www.paulhillson.com

THE SUBTLE SHIFT

The Fundamental Mechanism
Paving the Road to the
Extinction *of* Humanity

CONTENTS

	Foreword	
Chapter I	Introduction	*1*
Chapter II	The Subtle Shift	*20*
Chapter III	Thought and Thinking	*73*
Chapter IV	Reality	*91*
Chapter V	Time	*113*
Chapter VI	Dualities	*124*
Chapter VII	Language and Communication	*150*
Chapter VIII	Life vs. Living	*159*
Chapter IX	Disharmony	*169*
Chapter X	Parrots and Puppets	*175*
Chapter XI	Religion	*180*
Chapter XII	Final Words	*187*
	Author's Message	*197*
	About the Author	*198*
	Other Books by this Author	*199*

Foreword

When I agreed to be the copy editor for this book, I had no idea as to exactly what I was agreeing to, and how it would impact my perspective of the human condition. I questioned the author as to how the idea for this book came about, and he said, "this book didn't come about because of anything it seems I had done in my past, but in all honesty, it came about in spite of it." That reply in and of itself was enough to stir my curiosity, so, of course, I took the job.

The Subtle Shift presents a fresh perspective on humanity and the human experience that I honestly had never considered. While it is very intriguing, it can also be quite challenging in that it shines a light on one's ephemeral status as a person. Some themes do appear to be repetitive and redundant, however, by the time I completed reading the book I concluded that this was a necessary device employed by the author. At various points during the process of reading this book I caught myself dramatizing the very attributes that are inherent to the mechanism that is the central focus of these writings. The simple act of reading this book allows one to observe how the mind seems to function instead of getting lost in one's mental dramatizations.

After reading this book I came away with a profound sense that life is infinitely more practical than it appears to be on the surface. It seems that it is our thoughts about life that make it appear more complex and convoluted.

Von Lerma – *Copy Editor / Proofreader*

CHAPTER I

INTRODUCTION

Is humanity on the road to extinction? Is our ultimate path to destruction unavoidable? Might there be a way to reboot the species and put humanity on a path to a more "balanced and sustainable future?" In the larger scheme of things, does it even matter? These questions may sound somewhat ominous and foreboding, but perhaps things are not the way they *appear* to the limited human senses. Certainly, the sub-title of this book *seems* to paint a grim picture, but after reading this book you may notice a shift has taken place in your perspective on humanity and the human experience. And you will behold by yourself and for yourself, that existence is far different than it *appears*.

If one stops and observes what *seems* to be happening on this planet, the signs *appear* to be everywhere: terrorism, war, climate change, over-population, crime, religious and political divisiveness, violence, cancer of all varieties, Alzheimer's disease, Coronavirus, Ebola, Swine Flu, SARS, AIDS, Zika virus, etc. Moreover, one can add to this already extensive list the rapid extinction of other species. If one were to speak on the basis of *appearances,* it's impossible to repudiate the horrific impact caused by the exploding population of the human species, as man's footprint has reached the far corners of the earth and beyond.

Speaking on the basis of what *appears* as the human picture, if one takes into account the onslaught of news imposed on the ordinary person daily, it *seems* only sensible that people often feel hopeless and overwhelmed, and hence, motivated to find a way out, or perhaps a comforter or poultice of one kind or another. It *appears* that people are so habitually preoccupied with a past or a speculative future, that labeling them human beings is not only an oxymoron, but a deceptive depiction of their *seeming* status. It *appears* that there are more methods of distraction than there are people, but all of these diversions are simply ways of trying to escape from the *seemingly* endless challenges, pains, and pitfalls of living on this planet. The average person is only exposed to a minute fraction of what's really occurring across the globe, but all one has to do is turn on the news, read a newspaper or scan the internet, and a sense that all is not well becomes obvious rather quickly. It *appears* as though we're getting more news than ever before, and from places and sources we've never heard much about in the past. However, if one were to browse the news on a daily basis and listen to the multitude of sources available in print, online, and on television, it's rather easy to see that it's really the same news regurgitated and reported from an increasing number of purportedly diverse viewpoints. Of course, our sources for news tend to focus on or push a negative narrative rather than accentuating the positive, and it goes without saying that they undoubtedly have their own inherent biases. Speaking strictly on a human basis, regardless of how much or how little they are the subjects of discussion in the media, it *seems* as though there are enduring issues that threaten our *apparent* existence as a species on this planet. Surely few would argue with the statement that the way things *appear* is not always the way they are, so while much *seems* to be made about one's daily experience of living based on *appearances,* it wouldn't be prudent to blindly accept what *seems* to be reported by the limited human senses. That being said, when writing this book, I decided that it would be sensible to meet the reader on the basis of the human picture the way it *appears* in ordinary day-to-day experience. It may not be reality, but it *appears* that the vast majority of us *think and believe* it is.

Given the *seeming* state of affairs on this planet, not only in the United States but also throughout the world, I decided to explore a subject that had been dancing around in my head for many years. Let me be perfectly clear, I'm not a scientist, a geneticist, a psychologist, or an expert on human behavior, nor do I offer or feel there is a panacea for the multitude of problems and challenges that *appear* to be facing humanity today. If one were to speak in terms of what *appears* as the human picture, the author is as much a product of society as anyone else. In other words, from a personal perspective I'm really no different than you. This you that you know and experience as yourself *appears* to be a combination of conditioning, genetics, DNA, indoctrination, neurochemistry, and conceivably there are unknown elements as well. I'm not going to speculate about such things in this book. The point being made here is that in a human context, whatever or whoever you *think and believe* you are can only be contemplated through that which has been provided by others, and as a result it can be deduced that you are — at least in a personal context — a product of society.

If one were to assess things based on human sense, in the early stages of our development the honored value of being unique, special, a distinct individual, is often stressed while at the same time it *appears* as though we are subjected to a process of indoctrination, instruction, and conditioning. Early on, family, teachers, friends, and society in general, begin to impose information and beliefs into us, and therefore, it's actually impossible to be an individual when everything has been provided by others. How can people be distinct individuals when the only way they can know about themselves is through the knowledge that has been handed down from generation to generation? Despite the contradictory nature of this scenario, it *appears* as though people strive to be different, and many of them even pride themselves on their unusual skills, qualities, and originality, but upon closer inspection one finds they are not so distinct, not so original. People *seem* to go about their daily experience of living, promoting and projecting what they earnestly believe is their own unique viewpoint, when in fact this so-called unique viewpoint was acquired and assembled with bits and pieces from numerous viewpoints. This demand to be unique or different *seems* to have become increasingly prevalent in so-called modern societies. Some people call it ego while others just see it as narcissism, and to be honest, it's probably a little bit of both.

Speaking on the basis of personal or human sense, very often it *seems* as though conflict is considered good, and that the drive to be different is actually a drive to separate and divide. The word *individual* means, "single, separate, isolated." Constantly seeking to evolve beyond its inherent limitations, the purported personal self creates a demand to be separate, distinct, and apart from others, no matter what it takes and regardless of the consequences.

Based strictly on personal or human sense, via sight, sound, taste, smell, and touch, it *seems* that there are separate people with separate minds functioning within the context of a material world. The entire mortal spectrum *seems* to take place in time that never stops passing. It is in this limited context that, as people, we *appear* to grow and learn our common language, go off to school, and have a variety of experiences. Based on what is reported by the human senses, we *appear* to experience a three-dimensional material realm of space, time, and dimension, and as each generation *appears* to arise and recede, more knowledge and experiences are amassed and then passed down to the next. It *appears* as though this process has been going on since the purported early days of humankind.

Speaking on the basis of *appearances,* no one would deny that it *appears* as though we have developed incredible technology and made astounding discoveries, but at the same time it *seems* we have created a greater chasm not only between each other, but also between ourselves and the world that *appears* to be around us. It *seems* that humankind is obsessed with drawing lines, separating, creating division, and constructing walls at every turn. Irrespective of where or how you live, it's no secret that the world has always had its share of turmoil and strife. It *seems* that we are virtually conditioned to accept this scenario as normal, and most of us feel like there's nothing that we can do about it. It *appears* as though this has resulted in an underlying apathy that is part and parcel to the human personality. If one were to speak on the basis of human sense, our society, indeed the world today, *appears* to be more divided and has more ongoing conflicts than perhaps ever before. Of course, within the human spectrum itself that *appears* to take place within a field of time, it can be said that there are more people inhabiting this planet than ever before in human history. Perhaps this is at least a partial explanation for our current dilemmas, but too often this rationalization allows us to bury our heads in the sand, rather than react and take immediate action. Irrespective of the countless inventions, discoveries, and advances, along with the demand to generate a *seemingly* endless flow of information and experiences, there comes a point when important questions arise that demand to be addressed. What exactly is it that divides us? Where is this path that we *appear* to be on leading us? And is there a solution, a sort of global solvent that will ensure the survival of the human species? But there is yet one more pertinent question. Does it even matter? After all, what passes as one's entire experience of living *appears* to move within the framework of a continuous flow of time, and in this context, that which *seems* to arise within the framework of time must meet its inevitable end. Of course, making this sort of statement regarding time *seems* to instantly present a myriad of questions regarding the nature of time, its validity or lack thereof. The subject of time is so important in the context of this book that the author has dedicated an entire chapter to it, so rest assured, the nature of this thing we call time will be covered in detail.

When it comes to what *appears* as daily human experience, various cultures and teachings suggest that it's akin to a dream. Sometimes called: mirage, cosmic illusion, dream, waking dream, mortal-mind, sensual-mind, sensing mind, holographic mind, carnal mind, trance. These terms are synonyms that *seem* to reduce or dismiss the entirety of what *appears* as the human picture. If, as some have suggested, the human spectrum and its universe have no basis in reality, then it *seems* any attempt to reduce or devalue it would instantly present a paradox, in that it really wouldn't even be possible to discuss non-existence, much less attempt to explain or diminish it. Obviously, the simple act of trying to explain the nature of non-existence would in itself *seem* to give some credence and status where none is due, as non-existence is not being, not present, not now, and thus not, literally. This *appears* to present a bit of a catch 22. Rest assured, the various terms mentioned here are not used with the intent to devalue or make light of one's day-to-day experience. The author is not trying to undermine or discredit the purported achievements of humankind. The synonymous terms used here simply point out that the way things *appear* is not the way they are. It doesn't mean that what *appears* as the human scene is bad or evil.

This book makes the case that the entirety of the human picture, and the universe it purports to sense or experience, is mental and not physical. It is the activity of a never present temporal or human mind. This never present temporal-mind thinks, or dreams, and experiences a world that it *thinks and believes* is reality. Indeed, this entire spectrum is a realm of belief. The word *temporal* means, "related to or denoting time or tense." It's synonymous with the word mortal or human. The word *mental* means, "relating to the mind." Simply put, this means when speaking of what passes as human experience and its world, it is all mind or thought. It's important to explain the nature of this *seeming* spectrum because the primary focus of this book, *the subtle shift, appears* to have occurred within it. To be clear, this temporal mental construct is the activity of a never present temporal-mind. So, it *appears* as though the author was faced with an impossible challenge, attempting to explain that which doesn't really exist, but *seems* to *appear* as one's human experience, indeed as one's world.

Human or personal sense *seems* to assume that we live in a physical world filled with separate physical objects. While this *appears* to be one's day-to-day experience as reported by the human senses, long ago scientists debunked the myth that the universe is simply an assemblage of physical parts and concluded that the entirety of the universe is not tangible matter, but invisible energy vibrating at varying frequencies. Why is this important in general, and in particularly, why is this important within the context of these writings? Well, consider what this means. It may sound implausible, but it means that the entire human picture and its universe would be nothing more than energy vibrating at different frequencies resulting in innumerable holographic forms. This means that the human experience and its universe as attested to by human or personal sense would be entirely mental and not physical, as there is nothing physical and never has been. The temporal-mind's sense of its world is all there would be to its world. It also means that the assumption put forth by human sense that there is something called a mind in a brain that is in a physical body is a false assumption. Furthermore, it *appears* that there is only one *seeming* never present temporal-mind, and the activity of this mind *appears* as the human picture and its entire world. Given that the human or temporal-mind is all measurable energy, its activity would be in a continuous state of reaction, *appearing* in the context of passing time, time that is never present. This is why it can be said that in reality, there is no such mind, and therefore, no human picture or physical universe. This is not a book on reality, but surely reality is not subjected to nor impacted by the *seeming* passage of time that is never present.

While this book presents the hypothesis that the entirety of the human picture and its world is mental, the reader is not expected to blindly believe this statement. Through a bit of sincere inquiry it's easily discerned that this is the case. In this book the word temporal is used because the activity of the purported human mind *seems* to take place within the framework of passing time. It *appears* that way, but this mind's thinking, dreaming, sensing, and imagining, *seems* to concurrently create the experience of time.

If one were to speak on the basis of what *appears* as the human scene, it *seems* as though the thinking mechanism resulting from *the subtle shift* has hijacked the entire show. While this is not a book on what is sometimes called the absolute, or reality, the ephemeral nature of this never present temporal-mind means that it can't be one's actual identity. It's a wannabe imposter, an absent interloper if you will. Absent because time is never present, an interloper because it *seemingly* runs the entire show, constantly feeding upon itself to increase in strength and momentum. Of course, it only *appears* this way to itself. The human or temporal-mind *seems* to be the observer, the observed, and the experiencer. It believes all that it senses, thinks, imagines, and experiences, is reality itself. The temporal-mind's sense of its entire world would be all that *appears* to be to its world. It's just thought, a mental construct. Take away the sensations of a physical world, and there is no world at all. Consider this for yourself, be honest, and see if this holds up. Let's say for example all sensations of your world suddenly vanished. What would remain? It's like saying, remove all that's not present in the first place, and what remains. What's left is what's here now. Principally, the point being made here is, remove the sensations of your world and abracadabra, for you there is no world. If not sensed by the temporal-mind, then there are no physical objects, no observable forms, no sense of a separate self, and therefore, no experience of a world "out there." Speaking on the basis of human sense, this is how one *seems* to experience daily living, and it's on this basis that it can be said that the entire human picture and its world, is a never present mental construct.

As the temporal-mind itself is vibration, *the subtle shift* is essentially a shift in this vibration. Given that it's all reacting energy with no one at the helm, a radical or unusual shift in this vibration would most assuredly have ramifications regarding what *appears* as the human picture, and one's daily experience of living. It *seems* that this book is being presented within a three-dimensional temporal framework, so it's only sensible to start on the basis of the way things *appear* in ordinary daily human experience. In other words, it *seems* only sensible to discuss things within the context of what would be *apparent* regardless of the reader in question. Speaking on the basis of human sense, few would argue that there *appears* to be something called human thinking, and few would argue that it *appears* to dominate one's day-to-day experience of living. The essential issues presented in this book pertain to the *seeming* temporal-mind, the thinker of thoughts, and its connection to what *appears* as the human picture. This book looks at the ephemeral nature of this mind, and the likelihood that human thinking went off the rails at some point, i.e. *the subtle shift.* Additionally, these writings look at the nature of the human picture in its entirety, its relative reality or lack thereof, and the conceivable extinction of humanity, at least as we ourselves have come to know and define it.

Perhaps the temporal-mind's world and its experience of it is analogous to a dream, but even if one were to speak of a human dream, it's not possible for any particular character to exit a dream, and it *seems* that such is the case for us, as at least in a human or personal context, we are part and parcel to this purported human picture that is not only a product of, but inseparable from this never present temporal-mind. Given the *seeming* troubles that *appear* to be part and parcel to humanity a reasonable question arises. Does this inescapable dream have to morph into a *seeming* nightmare, is this inevitable or is it preventable?

Speaking on the basis of *appearances,* what is ordinarily called human thinking is a mechanism and not a conscious entity. Only thinking thinks that thinking is of tremendous value. To the so-called human or temporal-mind whatever it senses, thinks, and claims to experience is absolutely real, and there is nothing beyond its sense of its world. The temporal-mind and the world it claims to experience *appears* within the framework of time that is always passing, and thus, everything sensed, whether a sight, a sound, a touch, a smell or a taste, is always passing on, moving, ebbing and flowing, coming and going, but never arriving to stay, never really being. Every thought, every emotion, every sound, smell, touch, taste, sight, is fleeting, transitory, never arriving, never truly being, never actually present. It's as though this thinking mind wants to be through a process of becoming, but it can never get there, as there is here, and here is now. The word *be* means, "exists or be present." The reader will notice that the definitions of little words like be, or is, are often repeated in this book. This is intentional as it *seems* they are so often overlooked or misinterpreted in one's daily experience. The now that is now is not found in time, as time would be the great pretender, *thinking and believing* that there's a point when it is not now. The so-called human or temporal-mind doesn't really operate within a separate field of time, it is time! And time deals or thinks only in terms of past or future, never can it deal with now, the present. It's simply not possible to think of now, that would require time. Thinking, projecting, and experiencing *appears* to be its only activity, yet thinking is never present, never here, now, so it has no fundamental foundation in reality. Reality, what truly *is,* is real now, not after more thinking about it. Thinking is a mechanical process that seeks to separate and divide as this creates the illusion of permanence to itself. The purported temporal-mind is mechanical in the way that it functions, and its only drive is to constantly think or dream, and project, as this ensures the continuity of its pretense. Thought is a powerful vibration, at least that's what it thinks. In one's day-to-day affairs it *seems* that thinking has a necessary and relatively valuable role, but it *appears* that a shift in vibration occurred, i.e. *the subtle shift,* and as a result thought is no longer content with being a servant but seeks to be master of the house and run the entire show.

Speaking on the basis of personal or human sense, the temporal-mind *appears* to have tunnel vision, ensuring the continuity of its pretense *seems* to be its only purpose. The repetitive mantras are all too familiar; one must think of an answer, think of a way out, think of a solution, think of a better way to live, think of a way to come together, think of a smarter approach, think of a different or unique method, or perhaps a whole new way of thinking is needed. Starting with and from the point of view of human sense, it *seems* there's no other way, as humans always look to how they've done things in the past in order to decide what to do in what they think is the present, and of course, thinking is never present. It's not possible to think of now, the present, as that would require time that is never present. The so-called human picture *appears* within the context of passing time and given that humans *appear* to be functioning within the context of this field of time, it's not possible for humans to be present because it's not possible for time to be present. So, any so-called person attempting to live more in the present or more in the now, would be an illusory wannabe thought personality that will never succeed. There is no now, or present in time, and no time in now, the present. The human tendency to put things off to a future that never arrives is not only irrational but also futile, as it's simply not possible for a future to ever be present. Certainly, this sounds illogical and pointless, but there is no conscious entity making these decisions. This is part and parcel to how the temporal-mind *appears* to function. It sees the future as holding all the cards in terms of fulfillment. The temporal or mortal mind and the present cannot co-exist, because the temporal-mind doesn't really exist at all, and the present is *all that is*. Only now, the present, is truly present. The word *is* means, "to exist, be present."

Speaking strictly on the basis of *appearances,* humans *seem* to ignore what is blatantly obvious because it's too immediate. In terms of the human picture, it *seems* that whatever we used that resulted in the mess we *appear* to be in today is what we're attempting to use to create a better tomorrow. There's a peculiar sort of insanity in this approach. It *seems* obvious that we're only going to get a modified version of what we already *appear* to have, but we continue on blindly pursuing the possibility that our fulfillment will be attained in a future. The temporal-mind is extremely proud of its ability to think or dream, and we as we know and experience ourselves are inseparable from this temporal-mind, and so when confronted with problems and challenges, unsurprisingly the thought arises, "we must think of a way out," or "we'll think of something." The point is that our response is without fail always rooted in thinking, and this *seems* perfectly normal because we *think and believe* that many of the things that we *appear* to enjoy in the human picture today we owe to our ability to think. Of course, that which is thought itself could never imagine the complete absence of thought. After all, who or what are you absent what you *seem* to *think and believe* you are?

In every field of human endeavor, we strive relentlessly forward pushing the envelope further and further with the fundamental goal of improving our experience of living, and making it more enjoyable, more peaceful, more fulfilling, healthier, and yes, even longer. Curiously, it *seems* that there are many amongst us who desire to live on as long as possible regardless of the cycles that *appear* to be intrinsic to nature. While we *appear* to live in a world where everything is always leaving, passing, shifting and changing, we concurrently *seem* to seek permanence in a future that never really arrives. Of course, as we are inseparable from thought, we *appear* to believe, act, and live as though we're separate from nature so there's bound to be a conflict there. It *seems* that because of our demand to live longer and longer, we just can't accept the fact that as each day passes, we as we identify ourselves are moving closer to the end of the road. The denial grows and the distractions mount, but the inevitability of our destination is never far off. As one *seems* to grow older, the temporal or time-mind can no longer take refuge in the future, so it must live in misery or seek to be as now as now is, and of course, this is an impossibility. As the inevitable end approaches it can *seem* terrifying to this never present thinking mind, as its only concern would be to ensure the continuity of thought itself. The temporal-mind and all that it senses, thinks, emotes, or experiences, and imagines, has no foundation in reality, but it *appears* as though it *thinks and believes* differently.

If one were to speak in terms of the human spectrum, it certainly *appears* that since prehistoric caveman days we have accomplished much, humankind has increased globally, and it *appears* that humankind takes a lot of pride in this fact. We have harnessed the power to wipe out life as we ourselves know and define it. Thinking informs us that we are more important than all the other species on earth. Indeed, thinking has allowed us, even enabled us, to dominate all the other species on this planet, and in numerous cases they have been forced into extinction prior to what would otherwise be the end of their so-called natural cycle. It is due entirely to our *seeming* ability to think that we have committed horrific acts against other species. Perhaps nature will impose a penalty that puts the human species in its place, but for the moment the war on nature continues. Because the human species is no more or less valuable than any other species *appearing* on this planet, and because as humans we are not really separate from nature, the war on nature is actually a war on humanity.

Like the blind leading the blind, often we act as though we are separate from nature, and many amongst us *seem* to *think and believe* that humans are more valuable than the multitude of other species on this planet and insist that humans have been placed on this earth for some grandiose purpose. If it were possible for people to stop and be ruthlessly honest, they'd have to ask if thought is acting as a servant or more like a master, serving its own agenda, superimposing a false reality over what is already a finely tuned relatively harmonious construct. Given that humans *seem* to be inseparable from thought the continuity of thought is the continuity of humanity. This being the case, is it really surprising that human sense values the human species over all the other species that *appear* to live on this planet?

Given the *apparent* state of affairs in what we ordinarily call our world, and taking into account the numerous ideas, concepts, - ologies and - isms that dot the human landscape, a question arises that begs to be asked. At a basic level has humankind changed, have humans as defined by humans themselves really evolved beyond their primitive urges and instincts? Within the framework of the human spectrum many of us *appear* to live busy complex lives, and our words, languages, and cultures, have evolved, our technology has progressively become more and more sophisticated, and our armaments are unquestionably more deadly. But can humans really be trusted with their own inventions, or are the same primitive impulses still at work underneath all the pretty words and phrases, psychological mumbo jumbo, and sophisticated technology? If one were to speak on the basis of *appearances,* America, for example, has been at war or involved in a conflict of one kind or another for the majority of its existence. That's a startling fact if you ask me. My friends always point out, well, that's the way it has always been. Sounds reasonable, but my reply is yes, that's certainly *appears* to be true, and it *seems* as if there's no harm in allowing for this dismissive attitude but given the technology at our disposal today, inclusive of an insanely powerful and widespread nuclear arsenal, it *appears* as though it's categorically insane. In recent times it *seems* there has been an added emphasis on "a build it smaller approach," and atom bombs that have extraordinary precision are already in the works. It *appears* as though our ability to wipe out life as we know and define it on this planet is irrefutable, yet many still cling to this notion that somehow, in some way, we are going to bring peace to this world by going off to war, and we repeat this approach over and over again, because we really see no other option, and in fact, we don't have the ability to do so. As stated, we as we know and experience ourselves are products of thought, indeed inseparable from thought, and thinking can't see anything. Human thinking is mechanical in its functionality, downright fascist in its behavior. Speaking on the basis of what *appears* as the human scene, it *seems* we are nothing more than conditioned automatons, having no freedom of action, or power of choice to impose our personal will. It's really a tough pill to swallow, and truth be told, whether this pill is swallowed or not is irrelevant, as our oblivion awaits despite what we personally *think or believe.*

If one were to speak of the human picture, when it comes to our inherent pull towards conflict or war, it *seems* we arrived at a point long ago where we stopped looking for another way. Then again, it may be reasonable to ask whether we ever considered another way at all, or even if we have the capacity to choose a so-called other way. Given the magnitude of the destructive weapons we possess, it *seems* obvious and sensible that all wars in this day and age should take place on paper. If it were possible for us to be honest then we'd have to admit that all of our efforts in diplomacy have been mostly fruitless. It *appears* that there are many who talk of peace, sign peace accords, and peace treaties, but at the end of the day, it *seems* an enduring peace eludes us.

As I write this book it's impossible to ignore the fact that speaking on the basis of the *seeming* human scene, there are events taking place around the world that give one pause and concern regarding where this is all going, and if there is any way to change the course of events that *seem* to be currently plaguing our world. Speaking in terms of the human picture, it *appears* that those who cause most of the stresses in the average person's day-to-day experience of living, are the same ones busy promoting expensive solutions and selling magic pills to alleviate those same conditions brought on by these very stresses. And the ones who instill fear and anxiety are the ones who assure us that they can protect us from harm. This has evolved into a dangerous game that we can no longer glaze over and continue to ignore. For all of our progress, it has become apparent that humankind may be progressing toward extinction. The entire human spectrum may very well be akin to a dream, a never present fleeting mental construct, but that prospect in and of itself presents an interesting question. Must the dream called humanity have to morph into a nightmare? In other words, is the entirety of the dimensional human picture and its world, essentially the activity of a fictional temporal-mind, destined to fade into oblivion?

There are those that will argue, and steadfastly disagree with the views in this book, and given what we *appear* to be — in a purported human context — it's perfectly normal, as it's part and parcel to our quandary as it correlates to *the subtle shift*. Whenever a book of this kind is presented within what *appears* as the human scene, there's bound to be those who *seem* to take issue with its contents. It *appears* there are as many finite opinions and perspectives as there are people, but putting all that aside for the moment, it has become exceedingly obvious that the greatest threat to humankind is humankind itself.

CHAPTER II

THE SUBTLE SHIFT

The Subtle Shift is a book about a divisive mechanism that *appears* to be inseparable from virtually every aspect of human experience, and how this out-of-control aberration may be paving the road to extinction for all of humanity. The human spectrum may very well be analogous to a dream filled with dream characters, but if one were having a beautiful dream — if it were possible — would one really want to be woken up? Speaking on the basis of this *seeming* dreamlike construct, as characters *appearing* in this dream it *seems* as though we are facing an enemy that is inseparable from ourselves. Humanity has been hijacked by a divisive mechanism that *appears* to be bent on making the dream that is humanity into a nightmare. Undoubtedly, it can be said that all dreams come to an end, but if the would-be dream is nothing but a vibration, then perhaps there is no definite expiration date for its conclusion. After pondering these pages and digging into the functionality of the imposter called the temporal-mind, and the divisive mechanism that it uses to ensure its continuity, the reader will possess a thorough understanding of just what we *appear* to be up against.

The basic premise of this book is that humanity has been placed on an irreversible path to oblivion, due to a shift in vibration, *appearing* in this world of holographic forms as a genetic mutation, a subtle shift resulting in a neurological disorder in the human species. *The subtle shift* caused what is ordinarily called human thinking to go off the rails and morph into a divisive mechanism that is not interested in being a servant but insist on being master of the house. Hence, the fundamental focus of these writings is *the subtle shift*, and how it *seems* to have put us on an irreversible path that ends with the total extinction of humankind. Of course, as already pointed out, anything *seemingly* born in time must meet its inevitable end, so perhaps the mechanism that I speak of in this book is part and parcel to the human realm and has basically functioned in a *seemingly* stealth like mode since the purported beginning of time. After many years of increasing in magnitude and momentum, this mechanism *appears* to have mushroomed into an immense unstoppable force that operates in such an automatic covert like manner that much of what people do actually reinforces and strengthens it. This mechanism is like an out-of-control computer with an unsolvable glitch, or it could be described in more blunt terms as an incurable disease that continuously replicates and feeds upon itself. Regardless of the depiction used, this thought mechanism has *seemingly* hijacked the entire game, and it *appears* there may be no way to perpetuate the continuity of the dream and avoid that which humans may fear the most, that of total extinction. In the coming chapters the reader will see why this fear is inherent, indeed inseparable from the so-called personal self that *appears* as a character in a mental movie.

Speaking on the basis of human or personal sense, what is ordinarily called the human body *seems* to function under a basic drive, and that drive is entirely concerned with survival. This drive *appears* to come to the forefront in what are ordinarily called life or death situations. In these scenarios, the human organism responds and knows what to do. Strangely, thought abandons us in these fight or flight scenarios. Often there's no time to think, as thinking *appears* to be too slow. The body, the *apparent* physical entity does whatever it can, like those stories one hears about where people are able to lift a car off of someone with a sudden burst of superhuman strength. If they had stopped to think about it, then they would have likely arrived at the logical conclusion that there's no way they could lift that car on their own. If thought is completely absent, then limitations are absent. In reality, thought is never present, always absent, but it is *seemingly* present to itself. Even though thinking *appears* to be mechanical, it *seems* to have an inherent mechanistic demand for survival as well. And certainly, if one were to speak of the purported human picture, thinking *seems* to have a functional value, and it plays a role in our ability to live on this planet along with all of the other species. Unfortunately, the relatively natural process of thinking where thought is acting as a servant to the human organism has transformed into a divisive mechanism. This divisive mechanism is like a terrorist that has hijacked what *appeared* to be our natural way of functioning. It doesn't matter if it's just a temporal or mortal mind, a dream, or an illusion, because to those that *appear* to be part and parcel to it, it's quite real. Attempting to convince a person that their suffering isn't real, that they are merely a character in a temporal mental movie, and that their experience of living is akin to a dream is not sensible in any context. For example, if you were thoroughly convinced that you were drowning, yet simple observation indicated that the water was only three inches deep, it wouldn't make sense to stand there and argue with you. Helping you out of the water, and then perhaps discussing the facts of the situation afterwards would seem not only practical, but also rational.

Essentially, the divisive thinking mechanism that has manifested as a result of *the subtle shift appears* to be at odds with the human organism, dominating, falsifying, altering, and repressing its natural way of functioning. I hesitate to use the word natural, but this vibrational construct *seems* to function in its own way. This *appears* to be an ongoing conflict in humans because the thinking mechanism seeks to control, dominate, and interfere with every aspect of human experience, including the human body itself. This mechanism *appears* to be programmed to ensure its own continuity at any cost. Given that the entirety of the human picture and its universe *appears* to be energy, and in this context energy can neither be created nor destroyed, what we call the body or the human organism, already *seems* to be living a continuity. Thought *seems* to believe that this *apparent* continuity means life. It's as though thought seeks in vain to know life, indeed capture life, and it has no way of doing so. Through the use of thought it *appears* as though humans have invented numerous lofty sounding terms, but it's simply not possible for thought to define or describe, what is often casually called life, much less grasp or give expression to it.

Within the context of this finite realm where we *seem* to experience our day-to-day affairs, we have defined what it means to be alive, and what it means to be dead, and it *appears* that this has resulted in a tremendous anxiety and fear around the subject of death. Consequently, in our daily experience of living it *seems* as though we live with death as a constant companion. It's a cycle that is at work in all aspects of our experience within the human spectrum. Speaking on the basis of *appearances,* it *seems* that all of us must eat every day in order to provide energy for the body. From a physical perspective, what we define as life forms live off of other life forms. In a weird way we *appear* to live in a world where the process of eating is one of consuming death so that one may live. Can this really be called *living?* Generally speaking, we are burdened with a persistent anxiety around the subject of death and dying. Humankind has caused a lot of unnecessary disharmony in this area. We kill for ideas, for sport, while other species kill for survival. One can only speculate but there doesn't *appear* to be any maliciousness in their actions. It *seems* we are trapped within the context of ideas, definitions, and concepts that *appear* to be of out own making; indeed, humans are part and parcel to these ideas, definitions, and concepts. After all, what can we know about ourselves without the information, ideas, and concepts we have about ourselves? In other words, we are what we *think and believe* we are. Speaking on the basis of the human picture, what has been identified and labeled as the human body is an organism that *appears* to be immortal, and upon close inquiry, when it meets our definition of death, there is still plenty of life there — at least within the context of how we ourselves have defined life. There is an exchange, a realignment of atoms, and new life forms spring forth from what we define as a dead corpse. This process is analogous to our understanding of energy and that it can neither be created nor destroyed, only transformed into another form. Oddly, it *appears* as though the thinking mechanism is attempting to duplicate this process, and the only way it can do so is through separation and division, which ensures its continuity. The divisive mechanism of which I speak, is like an uninvited guest that refuses to leave your home. There is a conflict that is always there, an ongoing war, and as a consequence, as a species, we have manifested and *seem* to function as dream characters in a dreamlike mental movie that *appears* to reflect this ongoing struggle. The body *appears* to function in its own way and if anything, what we call human thinking acts as a persistent nuisance. Of course, ultimately what *appears* as a physical body is not physical at all. Science tells us that it would be just countless numbers of rapidly moving atomic particles, but for the sake of this discussion it *seems* we must deal with this relative spectrum that is imposed on us in a practical and sensible manner. Perhaps we are all simply characters in a dream that for many of us has morphed into some version of a nightmare. It's irrelevant, as it *seems* that we have little choice but to navigate what *appears* in front of us the best we can. It *seems* that most of us live under an erroneous assumption that implies that we are limited humans having a limited experience in a limited construct, and that our date with oblivion is inevitable. The numerous speculative stories, beliefs, and theories about what occurs after we meet the human definition of death *appear* to be methods that we employ to assist in accepting our inescapable fate. In the end, they are simply tactics we use to come to terms with what we have no control over.

The striking similarity in regard to how the human body *appears* to function, and how human thinking *seems* to function, led to the conclusion that *the subtle shift* likely has its roots in the human organism itself. All of it *appears* to be the same temporal-mind. The body, thoughts, emotions, and the numerous other forms that *appear* within the framework of the human picture. This entire dimensional mental construct would be nothing more than energy vibrating at varying frequencies. There's no one at the helm overseeing these patterns of energy. This vibration is constantly changing and reacting, hence the forms *appearing* in the human picture that is itself constantly vibrating and reacting *appear* to change, because it's all energy. The point being made here is if one were to speak of *the subtle shift,* then one would actually be talking about a shift or change that resulted in what *appeared* in the human spectrum as a genetic mutation, that triggered a neurological disorder in the human species. It *seems* as though an irreversible event occurred that has put humanity on a path to extinction. While our *apparent* existence is rather spurious and at best ephemeral, this *appears* to fly straight in the face of conventional thinking and beliefs about our purpose on earth. The temporal-mind doesn't dream of the human picture and its world, the temporal-mind actually dreams as the human picture and its world. So, our so-called current path is a directionless path, as the temporal or time-mind is nothing more than never present energy that is constantly reacting, changing, and yes, shifting. One might say that in a certain context, the entire temporal-mind is constantly mutating, so it goes without saying that what it *seems* to project would have to be a mutation. Even what and who we think we are *appears* to be inseparable from this divisive thinking mechanism, leaving us to function as *seemingly* separate individuals, fearful and isolated, within a field of time that is experienced as a continuum. Fundamentally, from a personal perspective it *appears* as though we find ourselves dwarfed by the magnitude, momentum, and relentlessness of a machine that can't be stopped through any volition of our own. Having its roots deep within the human organism, this mechanism even *appears* to mimic the behavior of cells, in that it is incessantly going through a sort of thought mitosis, generating more and more offspring that *appear* within its own illusory spectrum as separate little selves. It's intriguing that thought in its drive to survive, its demand for continuity, has essentially duplicated a function of the human organism, and of course it makes sense if one considers how well it *seems* to have worked for this machine that we call the human body. For example, the human embryo grows from a single cell that then undergoes mitosis, or cell division, and then it proceeds to generate more cells. While this is not a scientific treatise on the behavior of human cells, it's fascinating how cells *appear* to function. The focus here pertains to how division corresponds to growth, and the perpetuation of the human species, and how the divisive mechanism resulting from *the subtle shift appears* to be almost identical in its functionality.

There are numerous research papers and studies regarding different theories in reference to when early humans suddenly changed and how this may have been the result of a genetic mutation. Speaking on the basis of what *appears* as the human spectrum, one could say that within the context of this *apparent* time construct, the human species is a relative newcomer. Researchers have discovered life — at least in the way we define it — that dates back billions of years. Scientists suggest that they've discovered fossil evidence of ancient microbial communities that date back some 3.5 billion years. While it can be said that there is rarely complete agreement on these sorts of things, scientists suggest that the earliest known fossils of anatomically modern humans date back approximately 200,000 years. One can only speculate on what purportedly occurred in early humans where suddenly there was the appearance of more advanced tools, cave wall drawings, ceremonies of some sort, a new kind of self-awareness, a clear division that hadn't been there beforehand. The entire subject of language and advances in human speech must be considered as well. As stated above, there is an abundance of data one can access online if one should have the urge to satisfy one's curiosity regarding these theories. I'm not a scientist so I will refrain from making any claims one way or another, and if I were to go off in that direction it would be a complete departure from the focus of this book. As stated, speaking on the basis of what *appears* as the human picture, humankind is a relative newcomer on the scene. Of course, we would like to *think and believe* that we are here for some grandiose purpose, but clearly there has been life on this planet — that meets our definition — far longer than we've been around as a species, and before we even gave it a label and a definition. Most people wouldn't argue with the idea that humans *appear* to dominate this planet, but there are researchers and scientists who have introduced other not so commonly discussed perspectives. Some say bacteria are far more dominant, as there are certainly more of them, but there are many other so-called life forms that could also be considered as contenders for being the most dominant on earth.

A few examples would be ants, mosquitoes, and beetles. The point being made here is that we're not as dominant as we often *think and believe* we are, and our self-aggrandizing behavior often leads us to make peculiar assumptions that lead to irrational actions.

Regardless of how we feel, what we *think and believe,* or what we say about our importance, it *seems* we must deal with this *seeming* relative reality as it is imposed on us. Speaking on the basis of human sense, one can argue *seemingly* forever about the nature of humankind and its *apparent* reality or lack thereof. The divisive nature of the human thinking mechanism virtually guarantees that exact scenario. *The subtle shift* is most assuredly the basic dilemma we are confronting today, as it is a threat to our *seeming* existence irrespective of whether this spectrum is a temporal dream, a mental movie, or whatever manner in which one attempts to describe it. Given what we *appear* to be within the context of this dreamlike mental movie, it *seems* that we have no choice but to confront this issue head-on with ruthless honesty in order to conclude once and for all whether or not there exist a genuine solution, an answer that eradicates the question and forever eliminates the demand for a solution. Of course, if one actually were to put an end to the demand for a solution, it *seems* one would — at least to a degree — slow down the momentum of thinking itself.

Is it possible to measure the magnitude of *the subtle shift* and the resulting mechanism that has hijacked what we ordinarily call human thinking? A rudimentary example may provide some clarity. It has been estimated that the human body experiences roughly 10 quadrillion-cell divisions in an average life cycle. One can only imagine what we're up against when this model is applied to the divisive thinking mechanism. If one's own demand to think in order to resolve the human dilemma only strengthens the divisive nature of thought, then clearly our options are limited. It *appears* as though we're all in this together, we're facing a crisis where *hope, faith, and belief* will never provide the answers we seek. They may *appear* to provide a bit of comfort, and perhaps some temporary relief, but that's where it begins and ends. In reality, given that this temporal mental construct that *appears* as the human picture, is not being, not present, not now, and thus not, it's depth and magnitude would be the same across the board. None!

What I'm about to suggest is simultaneously important and unimportant. From a *seeming* personal point of view, for most people it may *seem* to be extremely important, even urgent, but from an impersonal perspective, it's likely that what I'm about to present here has no meaning at all. If the survival of what we have come to know as the human species is of interest to you for whatever reason, then you'll be intrigued by this hypothesis. What I'm about to put forth may sound ominous or perhaps to some it will sound silly, it may even sound too simplistic. Whatever your particular response or reaction may be, I have no doubt that life — as we know and define it — will continue to move along in its own way, and what is presented here will stand or fall on its own merits.

Every person that *appears* on this earth, *apparently* living out his or her personal story in this dreamlike mental spectrum has exactly the same mission. It's entirely possible that there are a handful of those that this hypothesis doesn't apply to, but this would be rare indeed, and if this were the case the other 99.9% wouldn't even notice. So as stated, every person on this earth, with perhaps a handful of exceptions, has an identical mission. Not a single person that *appears* to be on this earth has any freedom of choice or freedom of action in regard to this mission. People don't *seem* to have any individual awareness of this underlying drive, but even so, they are all seeking the same destination, and they move along on their journey accompanied by *hope, faith, and belief*. Everything a person does is motivated by a demand that came about as a consequence of a vibrational shift, i.e. *the subtle shift*. Every single relationship, every goal, every want, every desire, every search, every dream, is rooted in a single mechanism. It's really astounding when you look at this hypothesis in the context of living on this planet given the complexity and variety of the human spectrum. If one were to speak of *appearances,* it *seems* we have an infinite number of labels designed to describe our drives and motivations, but they are all without exception rooted in the same exact divisive mechanism with one singular goal. This mechanism *seems* to function on the same basis as the human body's own survival mechanism, and it seeks to ensure the perpetuation of what is essentially an illusion, and it will defend and protect this pretense to the very end. The functionality of this divisive mechanism makes sense when looking at the human picture and the entirety of its universe in a broader context. In that the temporal or time-mind, dreams as the human picture and its world, means everything is the same mental dreamlike stuff. It's all just passing thought constructs, patterns of energy, and this would include what is ordinarily called the human body. Speaking on the basis of the human spectrum and its universe, it can be said that it's all energy, all vibration, and this is why it's sometimes called the universal mind. Some have falsely claimed this to be god, reality, or the infinite, because it *appears* to be all there is to what is mistakenly interpreted to be existence or reality. Indeed, it *appears* to be, it *appears* that it actually *is*. When it comes to this never present temporal dreamlike construct, to say that *appearances* can be deceiving would be a huge understatement.

The never present temporal or time-mind is the thinker or dreamer, and the human picture and its universe *appears* as its canvas. It doesn't think or dream of it, but as it. Essentially, the mental canvas is inseparable from the mind itself. If one were to speak on the basis of *appearances,* it does *seem* as though a shift in vibration, i.e., *the subtle shift,* has amplified human thinking in magnitude and momentum. This shift in vibration may have something to do with the rapid pace in the growth of information, and the radical changes in societies across the entire globe over the last several hundred years. There's nothing one can do as an individual by applying one's personal will to reverse this vibrational shift, indeed, the more effort expended in that direction by any purported individual would only result in strengthening the mechanism itself, because any so-called personal self is not your actual identity, it's who you *think and believe* you are, but not who or what you actually are. The personal self would be a fictional temporal-mind's self, pretending to be your personal self, so any effort to reverse the vibrational shift, i.e. *the subtle shift,* would only result in a deeper sense of separation, triggering an increased demand to reconnect or unite, and thus the demand to resolve this illusory division would only escalate in magnitude and momentum. This *seems* to be how the never present temporal or time-mind inherently functions in order to ensure the erroneous continuity that it experiences as its reality. And any wanting or seeking to slow down the divisive nature of human thinking is futile. You as you know and experience yourself are inseparable from the mechanism that resulted from *the subtle shift,* so if this mechanism goes, you as you know and experience yourself would go as well. Some would suggest this is an awakening of sorts, but in truth, there's no waking up from a dream for a character functioning within a dream itself.

If the thinking mechanism that is part and parcel to the temporal-mind were to vanish into oblivion, it *seems* there most assuredly would be a vibrational shift *appearing* in the human organism as a physical alteration right down at a cellular level. This shift wouldn't have any spiritual or psychological connection, so it wouldn't be sensible to attach any type of label or religious meaning to it at all. This is all a bit speculative, but if this sort of so-called happening occurred, then perhaps the desire to know what had happened would be entirely absent. There's no point in speculating along these lines, but it *seems* reasonable to conclude — in the context of the human picture — that there have been various cases down through human history where perhaps nature has initiated another kind of shift, perhaps a realignment. Given the nature of human thinking, if this hypothetical event where to occur, it's likely humankind would attempt to use such a person as a prototype to produce others of the same ilk.

The divisive thinking mechanism resulting from *the subtle shift appears* to be the perfect trap. If you as you know and experience yourself attempt to escape from this prison, this separate little you will only *seem* to get in even deeper, and only strengthen the walls of the prison itself. Essentially, for this little you, this personal separate self, there's no way out. This purported separate personal self never wakes up, because this little self is a product of never present dreamlike thought, and thus inseparable from it.

On the surface, in one's daily experience of living, it *seems* that there are many who desire to feel whole, worthy, and valuable, but speaking on the basis of the human spectrum, how does one feel whole in a world where all forms *appear* to be separate from each other. How can one feel worthy and valuable in a society where what is considered worthy or valuable is the result of what is admired, and what is admired is often aberrant, if not insane? This trap *seems* to be so clever and so stealthy it's no wonder it has been able to essentially function on autopilot, taking over and dominating every aspect of human experience. This *appears* to be the situation but given that purported human experience and the trap are one in the same, this trap doesn't have to be all that clever or covert. It can operate right out in the open, as it truly only *appears* to have presence to itself. The saving grace is that this you that you know and experience as yourself, this you that you *think and believe* is your identity, are part and parcel to the trap itself. The thought may come, how can this be a saving grace. It is stated that this is a *saving grace* with no religious or spiritual overtones of any kind. The point being made here is that your authentic identity is never touched by any "thing" or "form" that *appears* within the context of time, nor can it be trapped or threatened by a never present temporal-mind. Life itself is the only one being present here, now, as the present. This is not a statement one is expected to believe. This is not something one gets. This is what *one is*.

Only the self can know or behold its authentic self, as there really is no other self to know or behold it. There never was a separation. Never was there a fall from grace. Life is all, and never divided or separated from its own all-ness. There is no inside to all (life) nor is there an outside to all (life). Not life in the sense of our dictionary definition, but the totality of life, the totality that life is, life that *seems* to be prior to any words or never present thinking about life. In reality, there's nothing prior to life, as life is now, and now is not a time. There is no prior to now nor is there an after now. This means there is no such thing as an afterlife. While popular in some circles, statements like, "we are all one or everything is interconnected" fail to convey the magnitude and clarity of that which cannot be defined or described with words. Words being dualistic conceptual containers fail when discussing what we ordinarily call life, but descriptively one might say it's an alive, conscious, non-numerical infinity, an undivided wholeness. In fact, it cannot be defined or described. Whatever is said is not it. In a personal sense, no one can help you in this area. Not the author by name, not a single priest, rabbi, guru or guide, not even a Dalai Lama can offer you more than temporary relief, a poultice or a comforter. For this you that you know and experience as yourself, this personal secondary self, there is absolutely no way out, up, or in. One cannot become what one already is.

The entirety of the human picture and its universe consist of the same substance. That stuff or substance *appears* to be energy, nothing but rapidly moving particles. Because these tiny atomic particles are always moving, always in motion, never stopping to be, calling this so-called energy a substance is really giving it too much credence. In that everything in this mental spectrum is made of the same never present stuff, it can be said there is an illusory unity to what is interpreted by personal or human sense to be reality. In spite of all the evidence suggesting this is the case, as characters in this mental movie, it *appears* we go about our business living under the false assumption that we are separate selves with separate minds, going about our day-to-day affairs as though separate from everything and everyone that *appears* to be in our world. Speaking in terms of the human spectrum, it *seems* that one can easily be convinced into believing that this world is just a collection of separate forms. But if it's all the same stuff, then what *appears* as a bunch of separate energy forms are not really separate at all. It's all simply a vast energy soup! This immense vibration or energy soup is always randomly reacting, always moving, shifting, changing, and never simply being. Surely this never present dimensional spectrum isn't reality, but from a vantage point within the soup it certainly *appears* to be quite real. This explains the futility of looking for a way out or searching for an answer. Starting under a false assumption, from a personal separate perspective, the moment one begins seeking one *appears* to move deeper into the vibrational dreamlike mental construct itself. In this never present realm, the one seeking and what is being sought are one in the same. It's impossible to see through this illusion because there really is no illusion to see through. The word *illusion* means, "mirage, false delusion, appearance, or hallucination." The point being made here is that what *appears* as the human spectrum is not real, as only reality is real. Only what is actually being, what really *is,* can be said to be real, or reality itself.

This may sound a bit cynical, but there's absolutely nothing one can do starting under the assumption that one is a separate personal self, as the divisive thinking mechanism resulting from *the subtle shift* has placed every person *appearing* on this earth on a journey that is doomed at departure. There is no choice but to make this journey. Exactly what is the point of this journey that is inescapably an expedition to oblivion? Every so-called individual *appearing* on this planet is driven to reconnect with that which they *think and believe* they have separated from. Each *seeming* separate self has no freedom of choice, as each self is inseparable from an illusory temporal-mind that is dreaming as the entire human construct. The temporal-mind is nothing but vibration, and within this vibration it *appears* as though a vibrational shift occurred, i.e. *the subtle shift*. The demand to unite, merge, or heal the *apparent* divides within the human picture is rooted in the vibrational shift designated in this book as *the subtle shift*. The demand itself *seems* to create an unnatural friction that results in yet a deeper sense of separation and division. As already stated, we don't *seem* to have any conscious awareness of this drive to connect because what we call human consciousness is not consciousness itself, and reality itself *appears* to be buried under a bloated and insanely complex thought structure. It *seems* this has been interpreted in some religious circles as the so-called fall from grace, but this was a false human assumption, because it was erroneously interpreted on the basis of a divided perspective. In other words, this assumption was put forth and fortified by the thinking mechanism itself. It does *seem* as though there are those amongst us who sense this disconnect, this separateness, and just don't know what it is, and so they are left to speculate on why this feeling of separation, emptiness, isolation, and fear always *seems* to be lurking in the background. There is an unexplainable sadness or melancholy in many people that is undeniable.

In terms of the human picture, often you hear of those people who *seem* to have it all yet continue to complain that there is still something missing, and yet it *seems* they have no choice but to continue on in a futile search for a sense of wholeness. This "you" who you *think and believe* you are, this "you" that you know and experience as yourself will never live with a sense of wholeness. There is no "getting there" for this *seeming* separate little self. What *you* actually are has never separated from infinity, from life. Life is indivisible and never separates or divides into little separate pieces. If this is truth, what is it that makes this you that you know and experience as yourself feel separate, wanting to get back, merge, join, reconnect, with that which you *think and believe* you have separated from? Thought creates the illusion of separate selves with separate minds. Ordinary human thinking is inherently divisive, but *the subtle shift seems* to have taken the demand to separate and divide to new heights. What and whom we personally *think and believe* we are today is inseparable from the mechanism that resulted from *the subtle shift*. This shift *appears* to have been a shift in vibration, and when this vibration shifted the energy form shifted, and what we call a genetic mutation occurred that *apparently* triggered a deep-seated neurological disorder in the human organism, that is yet another pattern of energy. The entire human picture and the world it experiences would be nothing but energy in a constant state of reaction, it's always changing, shifting, and reacting. No one, no "body," is in charge of this activity. What the author presents here is a hypothetical theory because *apparently* there's no way one can test what is being said here, and then proceed to confidently make the statement that this is actual fact. *Seeming* separate selves that are inseparable from, and a byproduct of the mechanism resulting from *the subtle shift* would be the ones conducting the test, and still again, the walls of the illusory prison would only be reinforced. All human experience would *seem* to be never present temporal dreamlike thought that *appears* to always be moving, shifting, changing, passing, and continuously reacting. You as you know and experience yourself are not separate from this temporal mental movie. It's all, without exception, the activity of a never present temporal-mind.

The biological machine we call the human body *appears* to be nothing more than an energy field, a vibration — tiny atomic particles vibrating at different frequencies. Even when it *seems* that the body form is asleep it is moving or vibrating. This means that a human body can't be a human being because a body is never being. This can't be what you are, an entity that is literally non-being. Indeed, the word *being* means, "existence." It's not possible to be what is not being. What you are in actuality from a personal perspective one can only speculate. In a personal sense we will never know the real self, the only true identity, because in reality, the authentic self is utterly impersonal. Reality is certainty. This certainty cannot be bequeathed to you by another, as reality precludes otherness. In reality, certitude is what you are. You are not just a character in a passing, fleeting, dreamlike movie. In a manner of speaking, you really are what you seek, yet every *apparent* movement, physical, mental, spiritual, or otherwise, can *seemingly* feel like one is seeking away from self for that which self is. Reality is not a higher self but the only self.

The separate self, the imposter that you *seem* to know and experience as your personal self, is not separate from the thinking mechanism that amplified as a result of *the subtle shift*, so if you *seem* to live with a constant sense of separation, division, or isolation, it is the mechanism that is imposing its will, always seeking to run the entire show. Speaking on the basis of the human picture, this mechanism is so driven to separate and divide, that it has even separated you as you know and experience yourself from the human body. How often do you hear people say, "this is my body," or "how's your body doing?" There's a separation there, a duality, and it has even made its way into how humans *appear* to communicate. In terms of the human spectrum, where we *appear* to have been and where we *appear* to be going is directly connected to the vibrational shift, *the subtle shift* that occurred in the human species many millenniums ago. We as we know and experience ourselves, and how we *appear* within the context of the human spectrum are nothing more than the outgrowth or byproduct of *the subtle shift*. I'm going to take a moment here and emphasize that I fully understand that it can be difficult to confront and accept this material, and naturally I wouldn't be shocked if there was a tremendous backlash as a reaction to its message, but then this book isn't titled, "how to win friends and influence people."

Speaking strictly on the basis of *appearances*, it's possible that what has been put forth in this book can *seem* to provide some with a degree of comfort, but the world *appears* to be full of comforters and poultices, and frankly, the author has no interest in starting a business in order to make available another one. In a strange twist this book may be a catalyst for a radical jolt, an energy shift that has consequences relative to our day-to-day experience of living. There's always the possibility of there being a total collapse of the illusory prison, but not as a result of anyone's individual effort. If by some fluke the prison walls were to collapse, it *seems* only sensible that any and all questions regarding these issues would cease to arise, and all seeking would be completely absent, even in a *seeming* sense.

Undeniably, in terms of the human picture as testified to by the human senses, it *appears* we are living in precarious times. It *seems* as though there are people who like to diminish the magnitude of our current status as a species. This *seems* to provide a false sense of comfort and security, but if the aim is to ensure the continuity of this thought dream, then it would *seem* reasonable to conclude that living in denial would be a dangerous game indeed. The author is under no illusions that people can be any different than they *appear* to be, and in reality, because the human experience *seems* to take place within the framework of passing time that never stops to be, technically, it can't really be said that there are people *being* at all. As lines are drawn, divisions grow, and more conflicts arise, and this is reflected in societies all over the globe. For some people it all *seems* to become a bit maddening and trying to adjust to an overwhelmingly sick society is no longer possible, but the majority see no other option but to go along on their *apparent* path, relying on *hope, faith, and belief* for comfort and relief. This might sound rather grim, and perhaps you will label me a pessimist, but I'm really quite optimistic. I suppose if you had to pin a label on me, I wouldn't squabble with the term realist. I'm simply presenting things as I see them, and undeniably, even the way I perceive things is biased and tainted by an uninvited interloper, a divisive thinking mechanism that seeks complete domination.

Speaking on the basis of human sense, it *seems* as though we find ourselves in a bit of a quandary. Humankind *appears* to be simultaneously blameless and yet accountable for the many problems that *appear* to be currently confronting humanity. This means as individuals there's little if anything that we can do, but because we *seem* to live out our stories within this thought spectrum, we are accountable, as it *seems* we are the ones who built it, and we're the ones who together can destroy it. Yes, the divisive thinking mechanism that resulted from *the subtle shift appears* to be in charge, and in a personal context we are inseparable from it. This is where accountability comes into the picture. It's not an easy situation because although we are not necessarily to blame, we are responsible for the current state of affairs. Even if it *appears* as though we're not, it's only sensible that we take full responsibility so that we have the ability to respond. In terms of this mental construct there may be little we can do, but that doesn't necessarily translate into do nothing. Things may not actually be as they *appear* to the purported human senses, but if we're going to live out our personal stories within the context of this temporal mental construct and *appear* to have functional societies, a degree of common sense and practicality is essential. It *seems* there are exceptions to every rule, e.g., being practical isn't always sensible nor is being sensible always practical.

The reader is free to agree or disagree, believe or not believe, a single word contained herein. It's understandable that there will be a lot of disagreement, as this is the nature of the human thinking mechanism. Attack, disagree, argue, protest, criticize, separate, divide, draw lines, create borders, and build walls, this is how the human thinking mechanism functions. Some readers will agree while others will disagree, as some individuals *appear* to unite while others *appear* to divide. Some amongst us preach and promote unity, but it's important to understand that it's under a false assumption that suggest there is separation. This approach may look harmless on the surface, but at the end of the day it actually generates more separation and division. The demand to join what is not actually separate creates a friction, and results in the opposite of that which one is seeking. In observing this state of affairs, it can be reasoned that the demand to separate and the demand to unite have similar if not identical consequences. These consequences *seem* to support the would-be temporal-mind in its mission as an interloper, as it seeks to ensure the continuity of its pretense.

Speaking on the basis of what *appears* as the human spectrum, it *seems* as though there's another facet to this divisive mechanism that is even more deceptive. It *seems* there are some amongst us in positions of influence who attempt to impose models on others, and often they assert that it's important that we behave a certain way in our interactions with each other, maintaining that this is necessary if we are to experience a better world and have healthier societies. Look around and be honest, in terms of the human picture. Have these *apparent* attempts to control the masses, to force a change, actually benefited humanity as a whole? Have any of these –

isms or –ologies been effective when it comes to harnessing or slowing the momentum and divisiveness so prevalent in human thinking? In all of humanity, taking into account the entirety of its purported history, does it *seem* that there is a single religion, method, or so-called approach *appearing* on earth that has really delivered the goods? The answer to these questions is self-evident.

If one concedes and accepts that the human species, as labeled and defined through the use of thought, is not really separate from the whole of nature, it becomes *apparent* that there are certain aspects of this dreamlike dualistic spectrum that will always *appear* to be with us, but this doesn't mean that things have to go completely off the rails. By attempting to enforce the demand of what we *think and believe* we want, and by over stressing one side of a fictional duality over another, we *seem* to throw things out of balance, hence we end up on the receiving end of the exact scenario we were hoping to avoid. Very often we seek what we want while our attention is on what we don't *seem* to have, or what we deem to be absent, and as a consequence we tend to get what we didn't want in the first place. Frequently, we *seem* to lock in the aspect of a duality that we consider undesirable. It *appears* that we fail to see that which we are seeking to avoid can't be separated from that which we desire. Humankind *seems* to not have the slightest inkling that they are perpetually fabricating what *appears* to be contrary to their intentions. If you're not dramatizing this in your own story, as an observer, it *appears* as a type of insanity. It's akin to the parable about the man who complains that his hand really hurts, and his friend says, "maybe you should stop hitting it with that hammer." And he responds, "what hammer?"

This process is woven throughout our societies. It may *seem* different when observing cultures in different parts of the world, but the basic mechanical nature of the process is the same, because it's all energy, shifting, changing, reacting, and it's all totally random.

Personally, I'm not trying to convince anyone of anything, nor am I attempting to bring anyone over to a specific point of view. In the context of this temporal dreamlike construct, I *appear* as an ordinary man simply singing my song by putting forth what I have observed, and as stated, in a personal sense, I'm part of this mental construct as much as any other character who *appears* with its context. If there *appears* to be something in these writings that *seems* to resonate, and ring true, then it likely touched something beyond words, ideas, beliefs, and mental concepts. There's nothing to get, let the book do the so-called work. If one were attempting to get it, then one would be a seeker. All seeking is in a future, and it is seeking itself that ensures the continuity of thought, as the future never arrives, the search must *seemingly* go on.

From a personal standpoint I hesitated when it came to the decision of whether to publish this work or not, as it's sure to induce a degree of controversy, and it will more than likely initiate more thinking, figuring, and searching. In fact, it wouldn't surprise me if the reader reacted to this work by criticizing, protesting, ranting, and reducing it all to total nonsense. What *seems* to happen or not happen in a never present spectrum only makes a difference to those characters that *appear* within its context and believe it to be reality. These *seeming* responses and reactions can be considered to be comparatively normal given the abnormal structure of the human spectrum as it *appears* today. If you so desire, you can be alert to your own reaction and response to what it presented here. Speaking on the basis of the human picture, the divisiveness intrinsic to human thinking *appears* to be easier to observe in others than in oneself. One can observe someone dramatizing, effectively functioning as the mechanism by being alert to a person's need to separate, divide, defend, and protect. In writing and publishing this book, the last thing I'd want to do is add more fuel to a fire that already *appears* to be out of control. It's not my intention to motivate anyone to go off seeking solutions in a future that never arrives. I personally have no control over how anything written here is interpreted or translated by any particular reader. There is no clear road to take, no genuine solution to resolve the *apparent* dilemma called humanity. The entire human spectrum *appears* to be nothing more than a temporal mental construct. What is stressed in this book is that what *appears* to be a dream need not mutate into a nightmare.

Speaking on the basis of what *appears* as the human realm, the vast majority of solutions are methods employed to avoid doing something as an immediate response to a given situation, and if nothing can be done as a direct response then any and all so-called solutions become faithful servants to the mechanism itself, and only strengthen the demand to find a solution. It *seems* to be an endless circular pattern where one is left thinking and figuring, thinking and figuring, thinking and figuring, seeking a solution and thus strengthening the mechanism that s*eems* to underlie the one insisting there must be a solution. Of course, if one is seeking a solution then they've already surrendered to the possibility of seeing one. Seeking is always an act of wanting something in the future, and in the future it remains. All seeking *appears* to be a projection of thought about a theoretical future that is simply thought thinking about thought. This is not only peculiar, but also mysterious, in that personal sense *appears* to be quite comfortable projecting thought into an imaginary future that has no fundamental existence. There's not even the slightest awareness that it is the seeking that fuels the urge to search. Would anyone search if they didn't feel somewhat lost? It is the search itself that strengthens that feeling, and a sense of feeling lost fortifies the demand to search. Wanting, seeking, searching, it's all the same type of drive that involves becoming, and it would be more accurate if instead of labeling a person a human being, we use the label "human becoming," or perhaps more accurately, "human non-being."

If there is any possibility of being freed from this illusory prison, it boils down to one word, simplicity. Actually, and factually, we didn't personally create the prison, no one did, as it really doesn't exist. The one trying to get out doesn't really exist either. We didn't originate the divisive mechanism that resulted from *the subtle shift* but building the complex structure that *seems* to have manifested over time *appears* to be our doing, and we increasingly fueled its continued growth and momentum. The temporal-mind thinks or dreams as the human picture and its world, and it also is the one that experiences it. In other words, if we as we know and experience ourselves were freed from this prison, we would still be relegated to living in a fleeting mental construct. So, in a personal context, no one is ever freed in the truest sense of the word, because the temporal-mind has no way of freeing itself from itself, and it is this mind that pretends to be the personal self with a separate mind. And yes, it *appears,* at least in a limited context, as though we're simultaneously blameless and accountable. Every last one of us *appears* to be on this road and there's little sense in denying our complicity. This is a game with no real winners in that we all have the same destination, so it's only sensible to play this game the best we can, identifying the goals, purposes and barriers, and pretending as though we are in control, even though evidence and common sense says otherwise. We have invented concepts in order to define a winner as opposed to a loser. So often we fail to see that winning and losing are two aspects of the same spectrum, and there's really no way to have one without the other. There are so many layers to these dualistic relationships that we either don't see them, or we choose to ignore them. Often when people *think and believe* they've won it turns out to be a loss, and when they *think and believe* they have lost it turns out to be a win. For example, in sports a loss can plant a seed that ends up being the motivation for next year's winning season. This would be speaking from the perspective of this dualistic dreamlike spectrum where we *appear* to live out our personal stories.

There's the way things *appear,* or the *seeming* activity of the temporal-mind, and the way things are, reality. It *seems* as though humans are all imprisoned within the context of an insanely complex mental movie, where an off the rails thought mechanism *appears* to infiltrate into every aspect of human experience, but *appearances* can be deceiving. And as mentioned above, if there's any chance of being freed from this prison that *seems* to be of our own making it comes down to simplicity. Freedom from this prison is in the clarity that there's no one needing to be freed from a prison because there exists no prison. But have the prison walls collapsed? In all honesty, does it not still *appear* as though there are separate people with separate minds who are looking for a way out? Someone who is free isn't seeking and searching for freedom. Seeking freedom is futile, but if there were an organic shift where one was freed from the demand to be free, then perhaps one *appears* to live a different kind of dream, but this it *seems* can't be attained through one's own personal will. Speaking on the basis of *appearances,* purported human history is littered with a multitude of methods, therapies, religions, and psychological constructs claiming to have the answers. They all have their roots in purported human thinking, and in effect, they only strengthen the demand to search, and that in and of itself has the effect of making one feel lost. Feeling lost, what is one to do? One seeks to resolve this feeling, one seeks into a future that is nothing more than a projected thought, as there is nothing lying ahead called a future.

If one were to speak strictly on the basis of *appearances,* it *seems* there have been people from various walks of life that have spoken and written on the subject of thought and human thinking. Many have come along and expounded on their observations, interpretations, and theories, so the potential impact of this material is not in the words, ideas, or concepts presented. Whatever results from this material is beyond any *seeming* individual's control, and I trust one will see clearly why this is the case by the time one has finished reading this book. I make no personal claims that I'm putting forth anything new. I'm using the same words and tapping the same well of human knowledge as anyone else who decides to publish a book. My genuine intent is impersonal and unconditional, but there is a sense that there's always a possibility that something beyond words, concepts, and ideas can initiate a shift from a place not rooted in human thinking. I honestly don't know if anything here has any value at all. In fact, I lean more toward the idea that it doesn't mean a damn thing, and in reality, it offers nothing in the way of genuine help to anyone. I will state that there is a certainty that says, *the only real help there is, would be that which precludes there being anyone, anywhere, at any time, in need of help.*

To be bluntly honest, I really don't see where there is any chance of anyone, anywhere, at any time, waking up from a dream. In fact, I don't see where there is any such thing as awakening from a dream, period. Even in the human scene, when you awaken from a dream, the characters that *appeared* within said dream never wake up with you. You are left with a sense that it never really happened. It was a dream and nothing more. While in the midst of a dream, the alleged characters never wake up to the fact that it's a dream. And when the dream *appears* to end, the characters are nowhere to be found. If this mental vibrational construct completely dissolves, what would be left no one can really know in a conceptual context. Perhaps something distinct, something like a rare flower may emerge. If you personally *appear* to be here, that would be thought itself, but in this hypothetical scenario thought would function more as a servant rather than an impostor seeking to control and dominate. I don't want to speculate in this area, as it may be exclusively unique to the organism in question. It *seems* sensible, indeed logical, that the obsessive wanting and seeking would be absent. This can be described as the end of the illusory search. There would no longer *appear* to be a fictional self, a little me, attempting to substantiate and superimpose its fictional narrative over whatever is there absent one's thoughts about it. Actually, it's impossible to really know exactly what would be left. It's like trying to describe what living and interacting with others would be like if *the subtle shift* had never occurred. This shift *appears* to have happened within a field of time, but in fact, it would be all there is to time. As time is when it is not now, now, reality or *what is,* has never shifted. *Only now is!*

Looking out from a personal viewpoint *seemingly* rooted in this fleeting mental construct, this never present temporal dream, it's not possible to know anything about life other than how we define and describe it within this conditional context. We can only theorize and speculate. The divisive thinking mechanism that resulted from *the subtle shift* is responsible for this *seeming* identity that is separate and apart from others, from nature, from the body, and from life, as humans *appear* to experience it on a daily basis. Certainly, you've heard someone say, "I have a life," or "get a life," or "it's my life." In this strange scenario, it *seems* as though they have separated themselves from life, as they *appear* to *think and believe* life is something that one can possess. Living as division, in a divided state, if that can be called living, it's only sensible that there would be a propensity to divide even further. If one comes from a perspective that implies one is separate from life, then can it be said that one is alive? Collectively all these beliefs, theories, concepts, ideas, mentation's, interpretations, and translations, have no life in them, as they are never present, they *appear* as passing thought constructs that are not being, not present, not now, and thus not, literally. That-which-is-not, is non-existent, not around anywhere.

The immensity that is life cannot be reduced to a word concept. It *seems* whatever is said about life, is not life itself. Even pointers are poor substitutes for the actual. When it comes to conveying the nature of reality it *appears* as though each of us must not only walk our own path but also pave it. I can only state with clarity that there exists a certainty that is impossible to transfer to another. It is up to one to meet with one's own certainty. It can never come from another, because life or *what is,* precludes there being another. I'm not asking you to believe in a personal viewpoint. One can use words to clumsily describe the nature of life, but words are never really accurate in this context. Collectively, all the ideas, concepts and interpretations about life are conceptual and not the actual. Life is now, so only now is alive. There is no prior now, no after now.

Undoubtedly, one can argue about the nature of humanity, what it is or isn't *seemingly* until, as they say, "the cows come home." But on the basis of human or personal sense, it does *appear* that humankind has arrived at a pivot point. As a result of what has been called in this book, *the subtle shift,* it *seems* as if an out of control thinking mechanism is attempting to dominate all facets of human experience, in fact, it pretends to be the human experience, and it's always seeking to increase its illusory power and influence. Does this mechanism possess any redeemable worth or workability? Can this mechanism be employed to solve the fundamental problems facing humanity today? Commonsense dictates that if human thinking would solve our problems, the world as it *appears* today would be free. All of our efforts to correct or improve the purported human mind only result in an increased sense of dualism. And when it comes to the many dualities that are part and parcel to every aspect of our experience, it *seems* there is a deeper and wider sense of separation between the two aspects that are intrinsic to each dualistic construct. The fundamental dualities one *seems* to encounter are as follows: *birth and death, happiness and unhappiness, health and sickness, wealth and poverty, love and hate, security and vulnerability, peace and war, success and failure, compassion and cruelty, pleasure and pain, hope and hopelessness, freedom and imprisonment.* These dualistic constructs that touch on every aspect of our experience of living within what *appears* as the human picture, *seem* to be complementary relationships that have gone awry due to the interference of thought.

If one were to assess things based on what *appears* as the human spectrum, it *seems* that many people relish their freedom of choice, but they often fail to see that they don't have any freedom from having to make a choice. When it comes to these dualities, each one is like a typical coin, once we *seem* to pick up the coin, instantly we must embrace both sides. In each case we can't have one without the other, yet we attempt in futility to accentuate one side of what *appears* to be a complementary relationship, and this only results in more complexity and conflict. And then we react and respond to what has manifested as a result of our own interference, and the vicious cycle continues. There's a peculiar madness to this approach because we really *think and believe* we're going to solve our problems through a process that only results in the *appearance* of more problems. Often, we feel it's working, but in fact the new problems act as a buffer, or at the very least a distraction from the initial problem, and in the end, nothing has been resolved, and after centuries of employing this approach one can only imagine the level of complexity that has amassed. This intricate web we have woven within the framework of this temporal vibrational construct *appears* to continue to feed upon itself, increasing in magnitude, momentum, and complexity.

Speaking on the basis of human sense — this world where it *appears* that we live out our personal stories — it *seems* as though we've reached a maturation point, we're on the receiving end of a constant and consistent barrage of beliefs, ideologies, methods, and theories, all of which have their roots in thinking, and it *seems* there is no end in sight. Given that the inherent nature of thinking is divisive, all of these approaches without exception are divisive, and so failure is built into the structure of these approaches right from the start. As the problems multiply and the divisions are fortified, we are admonished that we must be cleverer with our thinking, we must outthink others, we must think better, think more, think clearer, think this way, think that way, think deeper, think smarter, whatever you do, think! Basically, we are told repeatedly that we can and will think our way out of our current problems. We are encouraged, even prodded to think outside the box, yet generally speaking we don't *seem* to have the slightest inkling that thinking actually is the box.

After all of the events that have taken place, after all these years we can't face that which is painfully obvious. Human thinking will not rescue humanity from its ongoing problems.

Thought is a divisive mechanism that is only interested in the continuance of thought itself. It *appears* that if thinking completely ceased, the entire temporal mental construct, what we have come to know as the human picture, would have no status to itself at all, not even in a *seeming* sense. If anything were to manifest as a result of our thinking that even remotely looked as though it worked, it is thinking itself that confirms and points to this result in order to solidify its place as master of the house. If something happens to manifest that implies it didn't work, then thinking instantly kicks into gear, justifying, rationalizing, pointing, blaming, dividing, etc. Given that you as you know and experience yourself are not separate from thought, and that thought is exclusively driven to ensure its own continuity, it only makes sense that you would continue on as a faithful servant. It *seems* that it can be no different than the way it *appears* to be within the context of this passing dreamlike realm. If there's a situation or a problem that demands a solution, then thinking is the mechanism engaged, and it is engaged by thinking itself. It can only be described as a type of mechanistic insanity where thinking is trying in futility to figure out thinking, solve the problems that are byproducts of thinking, and this in turn, results in more thinking.

Whether the calls ring out from religious, scientific, business, entertainment, or political institutions, the message is basically focused on a belief that says the only way out of our current dilemmas is to somehow think our way out. Every field of human endeavor spreads its particular variety of propaganda, but they all have a common theme, we must learn to think better, think more, think this way, think that way, the emphasis is always on thinking. It *seems* that this is the only thing those in positions of power can do when it comes to addressing our problems. They can see no other approach. Why would they? They themselves are really part and parcel to thought. Thinking can never consider the cessation of thought, or the slowing down of thinking as a panacea to the dilemmas faced by humankind. This would be akin to a sort of suicide. Even attempting to quiet one's thoughts is silly, as one is left using thinking, and many times the mechanism kicks back generating more mental noise and anxiety. When faced with *apparent* problems, the mechanical impulse is to think! No one is at the helm making a choice. It's not difficult to see the futility in this approach as for every solution we put forth we *seem* to generate ten more problems, because they're not actually solutions. If something can be done it must be done now and starting now there is nothing that must be done. The purported solutions that arise from our thinking and figuring are always in the future. The issue that *appears* to be in need of a solution is strengthened and fortified by the relentless thinking, figuring, and searching for a solution. This can *appear* to be an exasperating situation, as so often we are attempting to think up solutions to imaginary problems. In all honesty, you as you know and experience yourself will never arrive at a point where you're completely free of problems. This scenario is part and parcel to the never present temporal thinking mind. I don't want to speculate, but I will say that freedom from problems is always now, only now, and now is not a moment in time. Surely you can't be liberated from your problems in a future that never arrives. It can be stated that the future never arrives because it's a self-evident fact that the future can never be present. After all, it's the future. Having put your potential solution in a future, in a future it remains, and given that it's impossible for the future to be present, your potential solution remains as a potential solution, a conceptual solution, and not an actual one. That which we are attempting to resolve we embolden without the slightest awareness that we're the one feeding the monster. A mechanism is not aware. A thought is not aware. A sensation is not aware. A thing is not aware. An idea is not aware. A body is not aware. *Only awareness is aware.* Identity *seems* to be a crucial issue because it *appears* as though there's a "we," and "we" are personally attempting to function as something that we *appear* to be aware of, and not as awareness itself. At least on the surface, all this *appears* to be a game where thought is attempting to capture life itself, but because life is now, and now is not a moment in time, it can never quite get there. Thinking is time, it's a process involving change, therefore, it's not possible for thinking to experience changeless, timeless now, the present.

In all sincerity do we really desire that which might put an end to the growth of this machine? This is a challenging question in that the divisive mechanism is inseparable from what we are as purported individuals. It *appears* we are part and parcel to it, and at the same time stuck in an illusory world of ideas, concepts, and beliefs. Why would that which is essentially the mechanism itself want to put an end to the mechanism when its only interest is its own continuity, the perpetuation of its pretense? In other words, what *appears* to be experienced as I, me, you, we, us, them, would be the activity of the temporal-mind. It *seems* to create the illusion of separate selves with separate minds, perpetuating more thinking, more separation, more division, and more illusory separate selves. And this *appears* to result in even more conflict, more agitation, more friction, more information, and more of an intense desire to know, to understand, to figure it all out, or to look for a way out. Collectively, this mechanistic process reinforces and strengthens an already formidable mental structure. It's insanely predictable, as this process is like food for its survival. This repetitive activity just goes on unceasingly like an assembly line device stamping out parts, drawing more lines, creating more division, and reinforcing the illusion of separation. And there's yet another aspect to the mechanism that resulted from *the subtle shift*. It *seems* as though once the separation or division *appears,* it then shifts into another gear, defending and protecting in order to define even further the *apparent* divisions. Hence, if we speak in a broad view of the human spectrum, lines are drawn, borders established, flags are printed, and before long a sense of nationalism takes root. In addition, this is exhibited amongst the *apparent* inhabitants in what *appears* as the human picture where people *seem* to focus on differences, and not commonalities, often an "us versus them" mentality takes root.

The simple act of reading this book will *appear* to make some people feel unstable, threatened, and uncomfortable. The thinking mechanism that is the central focus of this book functions as though it is programmed to disallow anything that *appears* to threaten the game, because its continuity might be severed and the entire structure may collapse, but because it has no individual will, it really has no choice but to function within, and as, its own convoluted framework. The mechanism in question is not aware, not intelligent, not really present, not alive, not real, not true, not being, not now, and truthfully, it is not existent.

Speaking on the basis of what *appears* as the human picture, we have come to a point where it's necessary to confront the only reality we know with ruthless honesty. It's time that we accept the cold hard fact that those who have come along declaring to have answers to the dilemma that is humanity have all, without exception, failed us. If one were to speak on the basis of *appearances,* there have been countless complex ideologies that have proliferated down through the centuries that have been employed as control mechanisms, comforters, and poultices. It *appears* as though humankind has been practicing and preaching methodologies and beliefs for centuries. One would be hard pressed to prove that any of these approaches has done anything beneficial at all. It's really amazing how loyal people *seem* to be when it comes to complex thought structures that never deliver real solutions. Religious people go on about their daily activities as though these approaches have shown the results that had been promised. This reveals a certain lack of freewill. It *seems* absurd, but they fail to see that regardless of how much they beg, plead, admonish, and practice, humanity *appears* to move along functioning in its own way regardless of the rules, precepts, commandments, and admonitions. It certainly *appears* that stealing, cheating, murder, adultery, corruption, and greed, continue on unabated, with no end in sight. It *seems* that centuries of practicing and preaching have done little if anything to curb their growth. It's astounding that no one looks at the total failure of these utterly false beliefs. The alleged believers continue in their commitment to falsehoods with more dedication than those who visit the local nightclubs with regularity. Even the most passionate and intense allegiance to a falsehood does not make it any less a falsehood. As funny as it may *seem,* there's no real difference between a believer and a non-believer. They both live in bondage to a concept, and both either, overtly or covertly would love to bring others over to their particular point of view. The believer believes in belief while the non-believer believes in non-belief. Each relies on the other for their illusory existence. What would one be without the other? If one or the other goes, then each would lose their fictional status. It is yet another coin, as it *appears* as though one side would have no status without the other. This is another example of how the mechanism *seems* to feed on its own dualistic framework in order to ensure the continuity of its pretense.

Regardless of what side one *appears* to take refuge in, it can be concluded with confidence that all of our ideas without exception, in particular all of our intellectual and religious thinking, have done little to resolve the fundamental impulses that *seem* to be inherent to humankind. Indeed, all of these ideas, concepts, admonitions, and theories may have actually had the opposite effect. Our institutions, whether religious, political, military, psychological, or academic, have failed us. In spite of this *apparent* reality, when problems arise, whether they are racism, terrorism, sexism, nationalism, you name it, we have nowhere else to turn. Our ideas and mental constructs only *appear* to make things worse. As stated earlier, I'm aware that some people will label me a pessimist. I have never been a fan of labels, but if someone for some reason found it necessary to hang a label on me, then being labeled a realist sounds sensible. At the end of the day you have no control over how others see and interpret you within this conditional mental spectrum. People are free to call me whatever they wish. I always give the example of walking into a crowded room; half of the people really like you and the other half can't stand you, it really has little if anything to do with you. Often people draw conclusions and pass judgments based predominantly on assumptions so that they can place a label on you and put you in a container. It *appears* that there is something in this behavior that is inherently comforting for them because they *think and believe* they really know, and it *seems* that because of the deep sense of separation, the divided consciousness that is inherent to individuals, not knowing has the effect of making the illusory separate selves feel like they're swimming in a wide-open ocean with no land in sight. Basically, the fear that *seems* to be intrinsic to these purported separate selves is magnified, and this *appears* to leave many so-called individuals feeling unstable, uncomfortable, and insecure.

If one were to look at humanity as a whole, regardless of the geographical location, there *appears* to be a familiar thread that reflects the commonality of people all over the globe. People from all walks of life *appear* to have similar ambitions, and often these goals disguise an underlying goal that they are busy working toward. Depending on where and how one lives, it may come down to the essential necessities for survival: food, clothing, shelter. It *seems* there are others who seek a relatively safe environment, after which, other essentials come into the picture. But what of those who have aspirations above and beyond the essential things that *seem* to be necessary to live? The list of things people seek is vast and diverse: freedom, enlightenment, peace, health, money, success, longevity, power, attention, relationships, friends, experiences, and of course, material things, such as houses, cars, clothes, etc. Essentially, when one gets beyond the basics one seeks happiness, and virtually everything else falls under that umbrella. People *seem* to want all of these things because they *think and believe* that once they have attained their wants, they will have attained a sense of happiness and wholeness. There *appears* to be an assumption that implies one is the totality of that which one acquires. At the end of the day only an unhappy person would pursue happiness, and that's a trap that can never be resolved, as one is reliant on unhappiness in order to know happiness. This seeking for happiness, and all seeking for all that matters, ensures that the thinking mechanism is able to perpetuate its illusory status as master of the house. Wanting and seeking are one in the same. They provide the spark that initiates the thinking mechanism. This movement of thought creates a peculiar agitation. The way this mechanism *appears* to function results in a deep sense of emptiness, isolation, fear, and separation, and together these consequences ensure the continuity of thought, because we robotically attempt to resolve these issues through the use of thinking. As characters within this dreamlike mental construct we don't really have any choice, and in effect, this is one of the fundamental dilemmas facing humankind.

Our experience of living not only *appears* to take place within a temporal mental construct, it *appears* to be a mental construct that is not actually being. This construct *appears* to be a realm of time and duality, where our personal demands become the orchestrator of the opposite of that which we are demanding. A demand indicates one has *seemingly* started under a premise that involves duality, with lack or absence, and then one seeks to fulfill that lack or absence. In other words, one started with lack, and hence one's consequences must bear the fruit of one's premise. This means that in pursuit of our goal we are always manifesting what we *think* is its opposite. Even if from the perspective of others, it looks like we got what we wanted, to us it's just not it, it's not enough, and something is not quite right, something is missing. If one *appears* to start with lack, there is no amount of acquired things, friends, knowledge, or experiences that will ever fill that lack.

Assessing things based on personal or human sense, it *seems* as though humankind has created a multitude of rules and models regarding how people should behave within this spectrum. This has resulted in a stream of overwhelming consequences, and the domino effect is revealed in our societies today. We can argue about what we are, we can argue about religion, politics, science, education, equality, and numerous other topics, but we can't argue with love and hate, happy and unhappy, black and white, peace and war, up and down, etc. We can't argue our way out of these dualities. It *seems* people treasure these dualistic constructs because they *appear* to offer a freedom of choice, but sadly, they offer no freedom from having to make a choice. Due mainly to conditioning, we *seem* to make these choices, *thinking and believing* that they are our choices. We *appear* to over stress one side of these dualistic relationships, because it's a divisive thinking mechanism making the choice. No one is at the helm! The demand to experience only one aspect of these dualities has its roots in *the subtle shift*. Obviously, it's not realistic that one should expect to experience only one aspect of any particular duality, but the master of the house that is the thinking mechanism isn't concerned with what's sensible. It's not surprising that it *seems* we've had an explosion of violence in many parts of the world. Speaking on the basis of the human picture, the duality of peace and war is prominent in practically every human society. Seeking to live in peace by suppressing and repressing war creates a buildup that at some point has to give. Visible evidence of the consequences of this approach can be observed in cultures across the globe. Humanity and duality are synonymous. Humanity's foundation is division, so in this context, how can we live without conflict? It's really not possible, but it's only sensible to conclude that there would be far fewer conflicts if both aspects of this complimentary relationship were allowed to be as they *appear,* absent the continual interference of what humans *seem* to think, feel, and speak. All of our ideas about how we should behave, what we should be doing or not doing, and how we should treat each other, are obstacles and irritants when it comes to functioning within this spectrum in relative peace. As this mental construct *appears* as a duality, it goes without saying that a lasting peace, a permanent peace, is not possible. We have so-called peace agreements, peace accords, peace treaties, cease fires, etc., but in the end, no real durable peace.

This may sound peculiar, but it's simply not possible to avoid many of the things we condemn the most. If one were to speak on the basis of the human picture, it *seems* we have to kill just to live. It's required! The demand to only acknowledge one side of an illusory duality in every facet of living is a result of *the subtle shift*. By stressing one side of an *apparent* duality, thought ensures the perpetuation of the illusory duality itself, through the process of seeking and searching for ways to only have one aspect of what is an erroneous relationship in the first place. Speaking in terms of the human spectrum, no one can be happy all of the time, no one can be at peace all of the time, no one can be kind all of the time, no one can be giving all of the time, no one can be tolerant all of the time. It's not even rational that one should expect to do so, but given what we are, or dare, I say, what we *appear* to be within the context of the human spectrum where these dualities *seem* to *appear,* we have no choice in the matter, so conflict is inevitable. It's just not possible as the mechanism that has hijacked ordinary human thinking is in a continuous state of conflict with the human organism, and the conflict is reflected in how the temporal-mind experiences its world. It doesn't matter if it is illusory, imaginary, or whatever anyone decides to label it. It is still imposed on us in a way that *appears* very real to us, at least in a *seeming* personal sense because the never present temporal-mind testifies to its own illusory reality.

The divisive thinking mechanism resulting from *the subtle shift, appears* to be programmed to generate division. The propensity to give more weight to one side of the numerous dualities within the human picture is like being forced to gorge on your favorite foods without cessation. It's only logical to expect that at some point one would grow sick and tired of eating even their favorite foods. When our want-tos become have-tos we arrive at a point where we no longer want what we thought we wanted. This may sound somewhat paradoxical, but what we are is thought, and thought never really *is,* it doesn't really exist. We *appear* to be stuck in a repetitious thought matrix, and we can no longer afford to sugarcoat our current situation. More information will not be the solution to the human dilemma. More fluffy words and empty phrases will not liberate us. Generally speaking, we make use of words and language to buffer things, to soften the blow as much as possible, and it simply hasn't served us well. Sometimes if you just call it the way you see it, there's a response, a reaction if you will, and thinking is bypassed, an action is taken. Thought is too slow, so when it comes to life, the idea that thinking can help us is proposed by thinking itself. It can't help us at all. *What you think you are can't be helped, while what you really are precludes the need for help.*

Seemingly underneath all of our ideas, concepts, and fluffy words about life and how best to live, what we most desire is already present. In fact, it is the present itself, and the present is not moving along with time. It is now! Life is now, not what was or what is yet to come. Thinking can't capture now. Starting from a personal standpoint, it would be wise if we didn't waste our time and energy attempting to understand this thing, or dare I say, non-thing, we call life, but many of us do, and it *seems* as though we really have no choice, as it would be never present human thinking attempting to capture life, now, the present. It's really an impossible task, but it keeps on trying because its functionality is analogous to a computer.

Wanting and seeking are movements that only lead to more of the same. If one seeks one feels lost, and if one feels lost one seeks. It's a vicious cycle. It doesn't matter what one *appears* to be seeking. This is part and parcel to how the mechanism *seems* to function. Religious people make their claims, those involved in so-called New Age circles and metaphysics have built complex thought structures regarding the nature of the so-called beyond, other realities, and various so-called levels of consciousness. Many people talk of pointers and living in the now or being more present. It all can *seem* rather silly, as it inevitably ends up as some form of business, entertainment, or an instrument used in order to avoid functioning intelligently in one's day-to-day affairs. All of these so-called approaches have their roots in never present thinking! Words are inextricably tied to thoughts, hence, it's not possible to think about now. That would require time.

Because of the thinking mechanism resulting from *the subtle shift,* our impulse is to place a label on every-thing, to put every-thing in a conceptual box, and hence we have arrived at a point where we automatically reduce whatever and whomever we *appear* to meet on our path. Essentially, it feels safer to cling to the illusion that we actually *think and believe* we know. Most of us are caught up in what was and what may be, as within the context of time, it's not possible to just be. This now that is now can't coexist with thought that is part and parcel to time. Now is not a point between a past and a future. Now is not a time. These writings don't discuss or speculate too much on what is sometimes called the now. I'll simply state that now is not part of this illusory mental movie. Now is reality. In fact, these are simply two words for the one all-inclusive actuality that is *all there is of all there is.* Words can't really do the job when it comes to speaking of reality. Anything said is not it, even what I just said here.

If one were to speak of the human spectrum that *appears* within the context of ever passing time, it *seems* as though we are always seeking to become something other than what we are, so it is becoming and not being that dominates our personal experience of living. Many people from all walks of life are driven by their wants, and wants correlate directly to thinking, and hence, becoming. The mechanism resulting from *the subtle shift seems* to have led us down a rabbit hole from which it *appears* there is no escape. It *seems* this way, but essentially, it has led itself down a rabbit hole from which it *appears* there is no escape because it really isn't interested in escaping. If we are interested in perpetuating the dream that *appears* as humanity — and of course we are — it *seems* only sensible that we can't continue to ignore the dilemmas we *appear* to be facing without incurring some serious consequences, and the definitive consequence may very well be the end of the human species. In reality, to reality, it makes not a bit of difference, and of course, this is exactly what so-called human thinking isn't interested in hearing. It *thinks and believes* that it really matters, because it seeks to really matter, it seeks to be real, to really be, but it has no way of being through the mechanical process of becoming.

Reality, now, or life — one is free to use whatever word one takes delight in using — *is all there is of all there is*. Nothing exists outside of all, so there cannot be a gain or a loss regardless of what *seemingly* happens in the eyes of humankind. There is no reality in a dream and no dream in reality. Whatever life, reality *is,* it is *all that is*. In fact, as fact, it's not really possible to define or describe life. Life or now *is,* regardless of what any "body," or any so-called human thinking mind thinks or says. It's never about the words, although some words *appear* to be more difficult to conceptualize, the word now is not now itself. The point being made here is that when one *appears* to be discussing reality it's never about the words because that which one is *seemingly* discussing actually precludes words altogether.

Within the context of what *appears* as the human picture, our wants and desires are often in conflict with the natural movement of things. Everyone has heard of those people who talk about how you need to get in the zone or get out of your own way. There may be something to this. So often when we totally give up, things have a way of working out, and on the flip side, in our zest for an instant solution, it *seems* that when we invest heavy effort on something it *appears* to only magnify the problem. People have frequently told me that they have had an experience of being in the zone or seeing something so beautiful that they personally were not there, but the fact that they can look back on it reveals they were very much there. It's just a clever trick on the part of the temporal-mind itself. There may be a stepping back, but there can be no stepping out of a dream, not for the little me that you sense, know, and experience as your personal self.

If our interest as a species, as characters in this dreamlike mental construct is our own survival, then we must come to terms with our situation and acknowledge that it *appears* as though things are trending in the wrong direction, and it *seems* we have taken a wrong turn somewhere along the line. The issue isn't how or why this deviation occurred. Humans *appear* inseparable from a temporal thought movie wherein each and every purported individual is simultaneously blameless and yet accountable. Can anything be done? Should anything be done? There are no easy answers to this quandary. Despite all of our ideas, concepts, and theories, it still isn't possible to escape from this never present thought construct and make contact with life at all. No amount of research will assist us in this endeavor. This is why looking for the life particle is entirely futile. We look for measurable evidence, observable forms, or phenomena as proof of life that is itself not a measurable observable form, or phenomenon. Perhaps this is our collective lot, maybe we're destined to live as fictional separate selves in a finite fleeting illusory world of ideas and mental constructs. Perhaps for us it's not about life at all, as there may be no way to know anything about life, at least from our inherently divided perspective. Just attempting to grasp or understand life is a divided approach that will surely fail. Certainly, if one is going to speak of what is true, real, fact, one must start with that which actually *is*. This is irrefutable!

Speaking on the basis of human or personal sense, it *seems* that many of the issues we personally *appear* to struggle with are connected to how best to live on this planet together. It *appears* we're always busy trying to figure out what is the right way, and in this context, there are opinions too numerous to mention. Despite what has been *seemingly* said in the so-called past, life is not a paradox, but it is our personal experience of living that *seems* to be paradoxical. We have come up against a complex mess that *appears* to be of our own making that covers the full range of daily living: politics, science, education, relationships, economics, environment, entertainment, religion, etc. It *appears* to be a convoluted box we have involuntarily fabricated, and it's almost impossible to confront, and as a response to this we keep devising ways to avoid rather than confront, and as our pleasures become our pains, we seek in vain for solutions that only *seem* to generate more problems, and around and around we go. The consequences of this approach are complex and unfathomable, adding layers to what *appears* to be an immense thought structure, creating more divisions and conflicts, and a deeper sense of separation from others, and from the world that *appears* to be around us.

As pointed out earlier, it *seems* reasonable, one might say normal, that certain questions arise because it *appears* that once in a while a so-called ordinary person is struck by lightning — metaphorically speaking — and for some mysterious reason this person is able to recognize that something is not quite right, somewhere humankind went off the rails. And I'm not claiming to be that particular individual by any means. But every once in a while, nature, life, whatever word or words you choose to use, puts forth something that is distinct, something that can't be duplicated on an assembly line. This is an *apparent* insight that may sound like it could possibly provide a genuine solution to our current dilemmas, but in reality, the moment we engage these questions we reinforce that which is responsible for our inherent sense of disunion, and our desire to know, to understand, to figure it all out is engaged, consequently the momentum of thinking is magnified exponentially. This being our *apparent* quandary, it's worth taking a look at the mechanics of what is in operation here, and if there is anything that can be done in order to ensure our survival as a species. If we stick our collective heads in the sand, then surely, we may be doomed. As characters within this temporal mental movie — typically called humanity — the human species may be doomed anyway. Given that everything within the human picture *appears* to have been born in time, maybe humanity as a whole is predestined to a final exit like so many other species on this planet. Why should we be any different than any other species, because thinking tells us so?

Some parts of this book can sound ominous. In fact, I hesitated to write it but after some pondering, I decided that it really didn't matter. There will be some that read it and throw it aside, and others may find it interesting to a certain extent, and perhaps, just possibly, there may be an individual who is touched at a point that doesn't *seem* to have its origins in the thought structure that *appears* to be reading, interpreting, and translating the words in this book. What happens from there is anyone's guess. It's simply not possible to predict, much less make sense of a vibrational temporal spectrum that is in a state of constant reaction with no one at the helm.

CHAPTER III

THOUGHT AND THINKING

If one were to speak on the basis of what *appears* as the human spectrum, what is it that *seems* to separate and divide us not only from each other, but also from the world that *appears* to be around us? And why is it important for us to acknowledge this fundamental vibrational shift, *the subtle shift,* that *seems* to have put the human species on an undeviating path to oblivion? As stated in the introduction to this book, if one were to speak on the basis of human sense, it *seems* clear that humans are indebted to, and reliant on their *apparent* ability to think. It *appears* as though thinking has taken humankind far from its modest beginnings. The divisive thinking mechanism that resulted from *the subtle shift* is so relentless and so all pervading, that we turn to thinking for answers to everything under the sun. And given that we as we know and experience ourselves are not separate from the divisive thinking mechanism that is the fundamental focus of this book, there's no reason why we wouldn't. Strictly speaking on the basis of *appearances,* as an organism, the human species is not separate and apart from what we ourselves call nature, but human thinking says otherwise. We have even separated humankind from nature by definition. In one's day-to-day experience of living, it *appears* as though thought can act as a servant, but the mechanism that has manifested as a result of *the subtle shift* is an interloper that has transformed thought into a tool it employs to ensure the continuity of its pretense. This limited conditional construct, the so-called human picture *appears* as a world of opposites, of separate forms, and numerous egocentric societies that are defined by separation, division, and by dissimilar cultures, borders, and flags. Within this dualistic construct, is it possible that we can attain a lasting peace? Given our situation, where do we turn? Is there anything one can do to resolve what is intrinsic to and inseparable from one's ephemeral identity? Human or personal sense *seems* to *think and believe* that our only available option is to think of a way out, think of a solution. The temporal thinking mind expounds endlessly on the value of thinking, and thus, as products of thought we employ thinking in order to resolve problems that are themselves the products of thinking. All of our thinking is dialectical in nature, and this *seeming* process acts as a catalyst, and we *seem* to empower that which we are trying to resolve. If thinking is responsible for the illusion of separation, can thinking be employed to heal or unite, and in some miraculous manner, vanquish the sense of separation that *appears* to manifest in a multitude of ways throughout humanity? Is it at all possible that these divisions can be dissolved through yet another method of thinking? The answers to these questions are self-evident. One cannot build bridges with the same instrument that generated the chasms.

What is ordinarily called thought or human thinking, separates, divides, protects, and defends. This is what it does. This is all it does! This is absurdly obvious by just taking an overview of the *apparent* human condition. If one were to speak of the human spectrum, it can be said that it *seems* to function as a continuous movement that results in the illusory experience of passing time. The totality of the human picture *appears* to be entirely mental. In other words, it's all thought. Thought cannot resolve the issues that *seem* to arise as a result of thought. In fact, any attempt on its part to do so would only result in the *appearance* of more issues that *seem* to need resolving. And surely, thought has no way of disengaging from thought. To the temporal-mind, the complete absence of thought is unthinkable! This is why it's not possible to think about now, for the now that is now knows nothing of time. Speaking on the basis of the human picture, it *appears* as though we have reached a point in our purported evolution within this illusory mental construct that there are so many different descriptions, classifications, and definitions for people that humanity is buried, lost beneath a multitude of layers, word concepts, and mental constructs. Why? What is it that separates you as you know and experience yourself, physically and otherwise, from those around you, from the world that *appears* to be outside of you? Is it not thought? In fact, would you as you know and experience yourself even *appear* if thinking were to be absent? If what is ordinarily called the human body is sitting and *apparently* looking at something, are you personally there prior to thought? To be ruthlessly honest, you as you know and experience yourself *seem* to be completely reliant on the senses, so you have a sense of self, a sense of a body, a sense of a world, a sense that there is a "you" that is looking at something, a sense of time, a sense of others, a sense of dimension, etc. In fact, this you that you know and experience as yourself is completely dependent on sensations. These sensations are inseparable from the temporal-mind. It's all just a vibrational mental matrix. The entire human spectrum and its universe can be reduced down to one word, thought. The mechanism resulting from *the subtle shift* has enabled thought to elevate its *seeming* status. Speaking on the basis of human or personal sense, thought even has a grip on sensations. Thought is constantly interpreting and translating every sensation, every thought, every emotion. None of these sensations, thoughts, or emotions ever really arrive to stay, so before thought has had a chance to interpret and translate, they have already passed and have been replaced with more sensations, more thoughts, and more emotions. This continuous interpreting and translating — thinking — *appears* to provide the temporal-mind with the illusion of permanence to itself. There's not a temporal or time-mind and sensations, thoughts, and emotions. It's as though a dream is dreaming as the entirety of the human spectrum and its world. Thinking, sensing, emoting, dreaming, projecting, imagining, speculating, and *seemingly* remembering, all would be the activity — dreaming or thinking — of this never present temporal-mind. This never present time-mind deals only in pasts and futures, consequently it *seems* to sense its own impermanence, and seeks permanence in a future that never arrives. The point being made here is that in one's daily experience of living, all one *seems* to be dealing with is thought, and thought itself is never really present, it never really *is*.

As soon as the thought "I" *seems* to arise there *appears* to be a separate self, and the *appearance* of others with separate minds. It's all illusory, passing, fleeting thought. It's as though this little "you" *appears* and disappears many times throughout the day. Various things *appear* or are sensed that *seem* to reinforce the thought that there is a separate "you," a separate self. Can you as you know and experience yourself personally separate from thought? It *seems* that as soon as there is recognition, suddenly you are there, and there is you and that which you are looking at. And simultaneously, there *appears* to be space, the viewpoint of dimension, and therefore, time. There is separation! It's like a magic trick, or more accurately it's like an act performed by an illusionist, because it's not real, not reality, but you (thought) personally "sense" it to be so. In a personal context, if you are looking at something, some visible form, could you have any idea about what it is you're looking at without the information that *appears* to be part and parcel to society at large? Clearly there is no way. How could you look at something and have any ideas about it, thoughts about it, or the ability to describe or interpret what you're looking at without the accumulated knowledge available to you as a consequence of educational input, indoctrination, culture, and what is contained in the global database that is repeatedly imposed on you over the course of your journey? In this conditional, finite spectrum that we *appear* to live and function in on a daily basis, it *appears* there is only a movement of what is customarily called knowledge or information, and it's incessantly repetitious. There have been studies indicating that up to 95% of thought is repetitive. That sounds like a disturbing statistic, even a bit crazy if you ask me, but this persistent repetition *appears* to be necessary in order to maintain the continuity of our illusory separate identities.

In addition to the *seeming* divisions, and the divisions

within divisions generated by this imposter — the thinking mechanism itself — are the ensuing consequences. The illusion of separate secondary selves with separate personal minds has given rise to the *appearance* of many of the things that we *seem* to enjoy in our modern society today. It also *appears* to have given birth to innumerable complex thought constructs, and automatic responses that are not at all pleasant. If one were to speak on the basis of the human picture, one of the most common and familiar byproducts would be fear. It appears that there's a multitude of offshoots dotting the landscape of humanity, but the majority of them fall under the heading of what is ordinarily called fear. Even to label fear a byproduct or a kind of offshoot isn't really accurate, as fear *appears* to be inseparable from the *apparent* separate structure you are familiar with as your personal self. Fear *appears* to be at the root of the *seeming* personal identity that you know as yourself. And there are many manifestations that *appear* within the human spectrum that *seem* to have their roots in fear: war, terrorism, oppression, suppression, animal abuse, exploitation, theft, conquests, tribalism, nationalism, arrogance, human trafficking, envy, narcissism, hate, jealousy. The list can *seemingly* go on ad infinitum, but in essence it's just creating more labels, drawing more lines, and as stated earlier, all of this *seems* to stem from this illusion of separation that instantly results in an *apparent* manifestation we call fear. It's impossible to discuss humankind without discussing the subject of fear, as essentially the two are one in the same. In a personal context, fear is inseparable from what we are, and although we can often distract and divert our attention or thought if you will, from what is intrinsic to who and what we *think and believe* we are, it's always lurking in the background, waiting for the right opportunity; fear of not getting what we want or losing what we have; fear of getting what we don't want; fear of the unknown, or fear of the known coming to an end; fear of the dark; fear of the light and what might appear if it were to come on.

In terms of the *apparent* human spectrum, people *seem* to put a tremendous emphasis on safety and security. Many politicians use fear to control the masses, and naturally it works to some degree, as people are always seeking a way to alleviate their fears. The entirety of the human scene *appears* within a field of time, individuals who are part and parcel to humanity are born in time, and that which *appears* to be born in time must come to an end, must die. Of course, when one is younger it's easier to pretend, as that inevitable day of reckoning *seems* to be far off in a distant future, then one wakes up one day and it's much closer than one could ever have imagined. Certainly, there's always the possibility of one's personal story coming to an abrupt conclusion. Everything within the context of this time-based picture is born in time, so we all live with the threat that it can come to an unexpected end at any moment. People *appear* to habitually superimpose a tremendous amount of thinking that generates so many distractions and diversions on top of this impinging *seeming* reality that they're able to go about their lives working and playing relatively unaware that oblivion awaits. And one could say they almost have to do this in order to create distance between themselves and the obvious facts intrinsic to functioning in this mental construct, otherwise the *apparent* reality that is imposed on them would be crippling. All of this *seems* to be driven by fear. We fear for ourselves, for our families, our friends, our safety, our health, our job security, our future, our country, etc.

If one were to speak on the basis of *appearances,* it *seems* that by giving birth to a deep sense of separation, isolation, and fear, the neurological disorder that resulted from *the subtle shift* drives us to divide, draw lines, and build walls in every aspect of our daily experience, as we seek to protect who we *think and believe* we are. To be an individual is to be separate or distinct from others. It *appears* as though there are many people who proudly declare that individuality is of great value, yet upon closer examination it *appears* as though there is very little individuality in the human spectrum. Within this never present temporal realm, all of us are more or less parrots and puppets, products of society, and our job — if one could call it that — is to live in service to the pretense that is the temporal-mind itself, by assisting in perpetuating the particular society in question. All of us use the same words, we have similar or the same thoughts, and we use the same words to interpret our sensations, emotions, and experiences. Despite the society in question and regardless of the particular language or geographical location, we really have no choice but to use the knowledge and information that has been put into us in order to have any sense of who we are, or what we are. As people we *appear* to be trapped in, indeed inseparable from, an ephemeral dreamlike mental world of ideas, concepts, and complex thought structures, and no matter how hard we try, we can't get out. The more we try to get out, the more we enlarge and strengthen the thinking mechanism itself. It's a futile situation and it *appears* that there's nothing we can do. The thing that has *apparently* given us so much, that which we are all indebted and loyal to, could be the very thing that pushes the human species into extinction. Perhaps loyal isn't the right word because we really have no freedom of choice in this scenario. And what *appears* to masquerade as loyalty is actually predetermined by formidable forces that have been built up and carried forward over thousands of years.

The subtle shift I speak of in this book is responsible for a multitude of mechanisms and automatic responses within the context of what *appears* as the human spectrum, but there is a fundamental foundation on which this illusory scenario has been built. What you're about to read may sound a bit crude, but if there is any chance of confronting our current dilemmas, it *appears* that more complexity is not the answer.

The most prominent and observable dominant aberration on this planet *appears* to be the impulse to connect with that which we feel separate from. What is it that the vast majority of us feel separate from? Generally speaking, most of us feel separate from everything and everyone, including the entire world that we *appear* to experience as being outside of us. Every single attempt to satisfy this demand to join, merge, or connect, will most assuredly fail, and will only result in the intensification of the sense of separateness and isolation that *appears* to be inherent to humankind.

If one were to assess things based on what *appears* as the human picture, it *seems* that through the use of thought we have invented our own definitions for life and death. For example, when we talk about searching for life on other planets, or throughout the solar system, we are more often than not talking about looking for the type of life that fits our definition. Frankly, I can't see anywhere where life would be absent, but I'm not attempting to convince anyone of anything. Simply consider the issue for yourself. It doesn't make practical sense that anything could exist outside of life. Yet, from a human perspective, people generally see life in things or on things, whereas a broader more sensible view would be that life is all-inclusive, and things simply *seem* to *appear* within life. In reality, it is only inclusive of itself, or *what is,* as there's no way to include what is not, as it's not here, now, to be included. In other words, life is reality and the substance thereof. This may be the subject for another book but it's not the primary focus of these writings. The point being made here is that human or personal sense assigns power to ideas, things, or forms if you will, and in doing so humankind *appears* to define life as finite, limited, transitory, having a beginning and an end. From this finite perspective, one considers oneself to be a mere thing, hence, the acquisition and accumulation of things *appears* to be the game. In other words, a pseudo identity is built on the basis of things. As one ages, often things begin to leave, decay, and as a result this pseudo identity suffers because part of itself is leaving. It can *seem* scary as one pretended to be the totality of one's things, and now they are all going away for one reason or another. The truth is that neither the pseudo identity nor its so-called finite things were ever present in the first place.

Imagine being inseparable from the all-ness of life itself, *that which is all there is,* and suddenly there is a personal you, and instantly this "you" has a sense of being a self, separate and apart from the all-ness of life, the all-ness that is life. In a peculiar way it *seems* as though there are some among us who sense this separation as they go about their day-to-day affairs. On many occasions I have heard people express that although they have accomplished much of what they had set out to do, there is still something missing, a void deep inside. This *seems* to explain why so many of us seek to get back into contact with that which we *think and believe* we have separated from, and unfortunately this only *seems* to strengthen that which we are attempting to free ourselves from in the first place. This "we" spoken of here, would be part and parcel to the activity — dreaming or thinking — of a never present temporal-mind, and thus, this "we" will never succeed in its pursuit of freedom. This is why we don't *seem* to have any conscious awareness in regard to why we're doing what we're doing. All of our so-called reasoning is rooted in so-called human thinking.

Speaking on the basis of what *appears* within the context of the purported human spectrum, if as characters in this mental movie we were able to strip away all of our superficial motivations, our real desire would be revealed. For example, the search for happiness — as it *appears* to be part of this movement. We search in futility for happiness within a fictional dualistic framework where happiness cannot even *appear* to exist without unhappiness. And we just don't want happiness in a moment, we want permanent happiness, and notwithstanding the temporal realm that we *appear* to live in, we continue head-on with our pursuit, all the while knowing it's not realistic. Because our primary goal is inherently unrealistic, the pursuit of happiness itself makes us unhappy, as this unconscious pursuit ends up being driven by a demand to avoid unhappiness. There is an inherent madness to this scenario because it's like playing a game we know full well we can't win, but in spite of what we know we insist on playing anyway because we see no other option. In fact, we literally have no choice, and this is why it's not possible for us to see any other option, as our response is inseparable from a mechanism whose functionality *seems* to be exclusively relegated to securing and solidifying its own illusory structure.

Frequently, people *appear* to be grappling with their particular version of a past, or a theoretical future. For many, just the idea of the present is potentially too painful — it's not possible to be present in time anyway — and so constantly avoiding and moving between a past, *that which is not anymore,* or a future, *that which is not yet,* becomes a habitual and, dare I say, mechanical way of functioning. It *seems* that we really have no choice, as personally we are trapped in this never present temporal realm, and thus, we move with the passage of time, and because time never stops to be, we never stop to be. In fact, it *appears* that we are not actually trapped in it, as we are really inseparable from it. Labelling it a trap would imply that there might be the potential to break free, but in reality, there is no breaking free for any so-called separate or personal self *seemingly* functioning within the context of a never present mental construct.

Time *appears* as a movement that is never present, and in a personal context we *appear* to live in this time construct, so we are never really present, even though it *seems* like we are. It's literally impossible to be present in time. This *appears* to create quite a dilemma. Things get even more convoluted because we rely on our memory, and we use the past to obsess about the future. What's next? What's next? What's next? There has to be something more, something better, something bigger, or something more fulfilling, some experience or thing that will be more fulfilling than this one, and hopefully, it will fill this lurking feeling that something is missing. It *seems* that at least for some this can be rather vexing because we seek our fulfillment in a speculative future that never arrives. In this ephemeral temporal spectrum, fulfillment is always tomorrow, never now. Perhaps a dose of common sense would help to alleviate our condition, but as the old saying goes, "common sense is not all that common." The thinking mechanism has no interest in common sense, and it *appears* to be the one in charge, the one running the whole show, in fact, it *appears* as the entire show, but it *seems* that in no way does this restrict us from being able to throw out questions. And this sounds like a great idea, but who is it that poses these questions? It is thinking itself seeking to perpetuate more thinking. Of course, the mechanism is not a living entity, it has no consciousness, no presence, no reality or life of its own. It functions exactly like a machine. Obviously, things can go awry with machines, and on occasion they can self-destruct. In spite of how a machine *appears* to function, there's always the possibility of a malfunction or a breakdown of some sort. Speaking on the basis of the human picture, its *seeming* status and history, it *appears* as though humanity has reached a maturation point, and it might be sensible to stop and ask some essential questions. Will more knowledge give us the answers to our search for happiness? Will more and bigger computers save us? Will more studies or more research provide us with the peace and joy we so desire? Will more religion provide us with liberation? Will more force give us security? Will more acts of war and violence result in peace? Will more data give us certainty? It *seems* everyone is starved for real answers, genuine solutions, and so far, all of those claiming to have real answers have failed. Speaking on the basis of the human spectrum, every so-called person on earth *appears* to be different, but the thinking mechanism is operating exactly in the same way in every facet of its own spectrum, even if it's covered up with pretty word concepts, diversions, and distractions. Perhaps as stated, it is the mechanism itself asking the questions, but it doesn't matter, like a machine it must operate based on the random reaction of numerous patterns of energy. It has no ability to make choices. Who knows? Maybe the machine has a built-in expiration date. I really don't know the answer. In all honesty, I don't even have the slightest motivation to find out. That demand would only *appear* to result in strengthening the mechanism itself, thus, ensuring the continuity of its pretense.

Today, as I write this book, it *appears* as though we've reached a new summit in humanity, a new peak in human thinking. So many of us now live in a virtual world, spending countless hours on our personal computers, cell phones, or tablets. Many of our jobs actually require us to live in yet another conditional virtual world that *appears* to be within the human picture. It *seems* that we are trying to use technology to solve our inherent sense of separateness and emptiness. In recent years we have experienced an explosion in the popularity of social media platforms. It's not all that surprising that the primary focus of these platforms is on bringing together or uniting *seemingly* separate selves, making contact, connecting, or reconnecting. Technology *appears* to have done so much in terms of our daily experience of living and how we communicate. It only makes sense to employ the same tool that has provided us with so much in order to resolve the *seeming* dilemmas confronting humankind. Unfortunately, or perhaps fortunately, it *seems* we can do nothing to stop this divisive mechanism, as it has a legacy that goes back thousands upon thousands of years, at least as purported by the human picture.

It *appears* that the never present temporal-mind has proliferated over many centuries into an unimaginable complexity, and it *seems* to move along with ever increasing momentum. This *appears* to be its nature, and it cannot be changed. The more one attempts to do something about it, the more one empowers and strengthens the mechanism itself. The tremendous amount of effort we have invested in our futile attempts to reconnect with that which we think and feel we've separated from has only resulted in a deeper sense of separation. This in turn has resulted in an increase in divisions, conflicts, and tribalism. It's really an exasperating situation that we *seem* helpless to do anything about. There are those among us who *appear* to embrace this madness, perhaps some are looking for a way out of their own misery by doing the opposite of what they *think and believe* they should do. None of it is personal, it's part of the same one mechanism at work, and it's constantly at work in all of humanity. Indeed, it's the very foundation upon which the entire human scene is built. This so-called foundation *appears* within the framework of time, and therefore it moves along with the *apparent* passage of time. In truth, it has no foundation in reality, that which is eternally now. This may not provide much comfort to you personally, but reality is not personal. In this book the focus is on the divisive mechanism that resulted from *the subtle shift,* and how it seeks to dominate every aspect of humanity. It may not be real, and we may very well be living in a temporal mental spectrum analogous to a dream, but to any character in a dream it *seems* very real, as it is the only reality this purported personal self will ever know or experience.

If one observes what *appears* to be occurring in our world today, the evidence of the divisive nature of human thinking is virtually everywhere. It's in every field of human endeavor; in every aspect of our experience of living, and in every geographical location. No one is exempt. No aspect of our human experience has been spared: war, disease, terrorism, religious fundamentalism, political divisions, crime, racial conflicts, economic struggles, family disputes, confusion, moral ignorance, suicide, depression, anxiety; it *seems* the list goes on ad infinitum. Proof of the divisive nature of human thinking is evidenced in all walks of life: religion, politics, education, medicine, sports, entertainment, science, spirituality, etc. It *appears* that there is no safe haven. The momentum of this machine is *seemingly* unrelenting, and as characters in this temporal thought construct, we *appear* to be supplying an endless flow of fuel. Our insatiable appetite for knowledge only bloats and strengthens what already *appears* as an insanely convoluted and complex structure. Essentially, the person that we have come to know, the way we define what and who we are, is inseparable from this mechanism, so everything we have done, everything we do, want to do, dream of doing, and plan to do, is in service to this machine. And as this machine becomes more complex and convoluted, it is reflected in what *appears* as the human picture, and therefore, our daily experience of living. At the end of the day our deep sense of separation has left a void that can never be filled. It's not possible as the void and that which is attempting to fill the void, are part and parcel to the same one never present temporal-mind. They are dependent upon each other for their erroneous existence. Not your individual demand, but the demand itself to connect, reconnect, or merge with that which you feel separate from is what *appears* to deepen the sense of separation and division that *seems* to be inseparable, indeed, undividable from who you personally *think and believe* you are.

As humans we function with a divided consciousness, all human approaches start from a divided perspective, a fictional duality, and hence, they start nowhere, go nowhere, and lead to nowhere, oblivion! The need, desire, and demand to unite, join, connect, make contact, or merge with the all-ness of life, the all-ness that is life, *appears* to have its roots in the concept of becoming. It's an invalidation of the undivided nature of life itself. This is a fundamental illusory conflict. You are not what you think you are, and whatever you are is not a was or a will be, you are now, or you have no existence at all. As humans we find ourselves in a somewhat maddening scenario that *appears* to perpetuate and expand like a machine that provides its own power, and thus it never ceases, as it has no reliance on any external source. In fact, to the thinking mechanism itself there is nothing external, as it operates on the basis that it's the whole game, and therefore, it dreams as the entire show.

If you personally discuss this subject with others, you end up fortifying and strengthening the very mechanism that has entrapped whatever it is that may be there, lying dormant and waiting to bloom. Whatever is there that *seems* to be waiting to bloom, if allowed to bloom would still bloom within the context of a temporal mental construct, but that's not the point being made here. The thinking mechanism discussed in this book *appears* on the scene as an interloper bent on making what *appears* as the human picture into an unmitigated nightmare.

This book and its contents may come across as complete gibberish and touch a nerve in the complex mechanism that *appears* to be reading the information contained herein. As an author, that's just a risk I had to take. I'm not any different than the reader, and like anyone else I must deal with this *apparent* relative spectrum that *seems* to be fortified and corroborated by society or risk being vanquished from it permanently. In other words, one plays the game the best one can, or one will be ejected from the game. This *appears* to be an issue we all have little choice but to confront head-on, and each new generation is conditioned to fit into a particular society and to become entangled in the game. There are those characters that just can't seem to adjust, and who don't want to be forced into a predefined model, for them, their experience of living can *appear* to become their own version of hell. For some, adjusting to a profoundly sick society triggers an out-and-out meltdown, and they commit suicide, end up being institutionalized, or living on the fringes of society.

In the upcoming chapters, I'm going to talk a little bit about various subjects that will *seem* to help clarify the nature of the mechanism that is the central focus of these writings, but I won't get too far-off track because this book is essentially focused on one fundamental issue, and that one issue is *the subtle shift*.

CHAPTER IV

REALITY

Reality is not the word reality, as no word or words can really define or describe that which is utterly ineffable. Speaking on the basis of the human picture, whatever we say about it, is not it. From a limited human perspective, we are stuck with the conceptual and not the actual. Some authors attempt to use pointers but, in the end, anything said about reality, *that which is all there is of all there is,* is never it. It really can't be pointed to or at, yet it is right here, now. Reality is what actually *is*. Yes, it *seems* there is that which *appears* but what *appears* is never actually present, never being, hence, not what really *is*. The finite human senses cannot perceive reality, the infinite itself.

The never present human or temporal-mind thinks or dreams as the entire human picture and its universe, it is said to be never present because time is not something this mind not only *seems* to function within, but it literally is time. It's impossible to be present within the framework of time, and thus, time and that which *appears* within the context of time is never present. It *seems* to be present only to that which is also never present, sort of bearing false testimony to its *seeming* reality. Time will be discussed thoroughly in the next chapter so there's no need to delve to deeply into that subject here.

The temporal-mind would be just directionless energy reacting, changing, shifting, and flowing ceaselessly. The vibrations are always changing, shifting, so what *appears* as physical forms are not physical at all, but changing and shifting energy fields — holographic images. The entirety of the human picture and its universe would be the same never present energy vibrating at varying frequencies. In other words, it's always moving, never simply being. The entire human spectrum would be nothing more than thought, a fleeting, mental construct. There really is no we, no me, no you, no us, no them, no our people or those people, as it is thought that *appears* to generate all these conceptual divisions in order to ensure the continuity of its pretense. The temporal-mind is its world, and it would be this temporal-mind that is *seemingly* experiencing its world. What *appears* as a physical or material universe is fleeting, passing, never present, dreamlike thought patterns that never stop to be. This purported physical world is not actually physical, but mental. It *appears* to be experienced by a passing, fleeting, personal you, but this personal separate "you" is none other than the temporal-mind masquerading as a separate little self that you *seem* to know and experience as none other than your personal self. Humanity, inclusive of all the *seeming* characters would be analogous to a mental movie. It certainly *seems* to be real, but only to the characters that *appear* to function within the context of the movie itself. Thus, the way things *appear* is definitely not the way things truly are.

This *seeming* journey that humankind *appears* to be on has been a protracted process, but basically, since *the subtle shift* it *appears* as though humans have been very busy inventing numerous labels, beliefs, theories, knowledge, and experiences. Collectively, what *appears* as the human realm *seems* to be nothing more than a finite temporal dreamlike spectrum. It's all vibration, a limited, passing, shifting, fleeting mental construct. If this entire construct inclusive of its *apparent* contents: bodies, planets, trees, cars, houses, oceans, air, fire, water, etc., is nothing but thought, it *seems* only sensible that the never present temporal-mind would be somewhat resistant to the fact that it has no real foundation in reality. Yet another reason why there may be some amongst us who *appear* to have a negative reaction to the message epitomized in these writings.

If one were to speak in terms of what *appears* as the human picture, it *seems* as though there's been a long progression, a sequence of numerous events within the framework of constantly passing time. It *appears* that somewhere along the line on this illusory journey, the divisive mechanism resulting from *the subtle shift* hijacked what is ordinarily called human thinking. It *seems* as though this usurping of human thinking triggered a demand in the human species to separate, divide, and generate beliefs, theories, information, and experiences, that are collectively handed down from generation to generation. This mechanical process adds more and more information, and more and more layers to an already complex and convoluted structure. Within this mental movie this knowledge is passed down under the false pretense that implies that it's really possible to know. Often, no one really stops to check the credentials of those who claim they know, much less whether it's possible to actually know in the purest sense of the word. In essence, it *seems* we know what we've been told, what we have read, what we have heard, what we have acquired and experienced on the basis of the knowledge that has been put into us by society, culture, etc. Everything we know is relative, conditional, inherently limited.

So-called human thinking is neither good nor bad, it just doesn't really exist, as it is never present, never being. If something isn't here, now, not present, how can it be said to exist? Within the *seeming* picture that *appears* as the human realm, thought *seems* to have a practical value, but ordinarily it would burn up in the action process initiated by thought itself. One can only speculate, but it *seems* as though certain animal species function in this way. Perhaps there's a thought, an impulse that says hunger. The word hunger may not be there but the impulse, the instinct, takes over initiating a reaction. Perhaps even human examples *appear* that pertain to this context. Let's say it's late at night, you get off work and go out to your car, you're about to go for the door and suddenly you catch out of the corner of your eye a snake, in an instant a thought, "danger," the body slowly moves back. The thought is gone, burnt up in the action. The organism takes over as there's nothing more that thought can do in this scenario. Then you get to a safe distance, perhaps more thoughts arise. Thinking takes over, seeking to control all aspects of the so-called human experience, it shows pictures and creates fictional stories about what could have happened, and one senses fear and anxiety even though the potential danger has passed. Thoughts may arise, "wonder if this happened, perhaps I should have done this or that, next time I'll know what to do." It doesn't *seem* as though thought functions in the same way when it comes to less complex energy patterns, like with different animal species. In these cases, the machine-like linear process of thinking *appears* to be absent. Speaking in terms of the human organism, the energy pattern *appears* to be more complex, and thus, the functionality *seems* to be entirely different. If one were to speak of the human picture and its universe, it's all energy, vibration, and no one controls it or orchestrates how it functions. It's always reacting, changing, shifting, ebbing and flowing, coming and going. This means that there is no one that is in control of making decisions or choices. What *appears* within the context of the human picture would be the consequence of constantly shifting energy patterns that are not being controlled by a distinct entity. Indeed, the entire energy field — the temporal-mind — is not being at all.

Thought *appears* as a wannabe, and it wants to know, but that's not possible, as it can only *think and believe* that it knows, as its entire status and the world it experiences is inseparable from belief. In one's daily experience of living, where it *appears* there are separate selves, it *seems* as though all of this information and all of these labels have a practical value, as it can help us to facilitate communication. What *appears* as a never present human picture is a mental construct *appearing* as a conditional or relative reality, and the glue that *seems* to hold it all together is agreement. As characters in this thought construct, we *seem* to share common knowledge, and we project that knowledge, and it becomes our experience of the world. For example, we all know in various languages what a chair looks like, what its basic function is, so we share this agreement that anything that looks and functions like what we have defined and labeled to be a chair is accepted as a chair. It doesn't matter who you *appear* to be or where you *appear* to be located, it is fairly common knowledge, as it is the same one never present temporal-mind making the claim. This is the one never present temporal-mind bearing false witness to its own ephemeral status. In other words, thought is literally never present, but to itself it is permanent, and its world is real. It's always seeking to be through a process of becoming, but it never really arrives because it can't stop to be. The word *be* means, "exist or be present." If the movement of thought, this movement that is thought, were to stop, then it wouldn't even have a *seeming* status to itself, as it really doesn't exist, it has no presence of its own. In a practical sense, whether it be a chair, a car, even a body, it *seems* that someone came along, made up these labels and enough people agreed with them to the point where they are instantly accepted as such. This is all part and parcel to the activity of the temporal-mind. Speaking on the basis of personal sense, we are told what things are and what they are not. This process is applicable in every facet of human experience, and it must be said that it certainly *seems* to have practical value, particularly when it comes to communication. For various reasons, it *seems* as though we don't want to admit that we don't know what we're looking at. Whatever we *seem* to be looking at we've been told what it is by others. We really don't know. It *seems* we can only "know" within a conditional framework. But if it's conditional, can it really be said that we know, that it's true? The word *truth* means, "the quality or state of being true." It *appears* that when speaking of the human spectrum we are stuck with nothing but relative facts and relative truths. Again, we are relegated to the conceptual and not the actual, not reality, what really *is,* or truth itself.

In a practical sense, the process of inventing labels and definitions *seems* to make it easier for us to function in this world and in our particular society. It *appears* to have its place in this temporal spectrum, but a divisive thinking mechanism has taken this process to new heights. Today we have invented labels for things and creatures, that at least theoretically, were on this earth millions of years before we were here to invent a label. For example, scientist and researchers have come up with labels and semantics in order to differentiate between different species. Basically, they develop knowledge, acquire and invent information about each one, and then pass that along as though they know. But let's stop for a moment and look at a rudimentary example, within our own finite context humans speculate that the cheetah has been on this earth for approximately 10 million years. This means that for the majority of its *apparent* existence it had no name, no label. Humans come along, give it a label, and then suddenly that's what it is. Sounds a bit arbitrary if you ask me. If we could just be honest for a moment, we would have to admit that we really don't know what it is, but it does *appear* to be something quite extraordinary. Why is this specific example important? Well, it's important because first, it *appears* as though this can be applied to everything, and secondly, over time one can observe how we've used language to separate, divide, and place random value on certain species. On occasion, I hear people make comments after someone loses a pet like, "oh well, it was only a dog," but in fact we really don't know what it is, or quite honestly, we don't *seem* to know that it never really *is*. Within the context of the world we *seem* to experience, it's just an invented word concept that we placed on something we perceived via our limited human senses. The process of putting it in a container by placing a label on it makes us feel comfortable in our relationship to it, and it also helps us to facilitate communication. This approach isn't exclusive to just other species, we even do this sort of thing with the human species. I can always tell when a certain group of people is in trouble somewhere in the world, and it's when those in positions of power start labeling them, "those people." It makes it a whole lot easier to start the bombing if you debase people first. The point here is that thinking is essentially knowledge, and knowledge — language — separates us from life as we *seem* to experience it in this finite context, and we use thought to distance ourselves from our transgressions against other people and other species. It's staggeringly complex and convoluted, as it has been *seemingly* constructed and fortified for centuries. When speaking of the human picture as reported by the human senses, it *seems* as though we have employed thinking in order to change, fix, and improve our experience of living based upon our interpretations and assumptions of what is needed and wanted. Additionally, we have employed thought in futile attempts to understand life, and all to no avail. This may very well be one of the main pivot points where things went off the rails, and it *seems* there is no going back.

The temporal-mind *seems* to have an incredible ability to dream up, conceptualize, and project things that become incorporated into what *appears* as our agreed upon dreamlike reality. The divisive nature of thought has made this even more layered and convoluted than it *appears* on the surface, because as characters within this dreamlike mental construct we experience our world through our particular thought structure, and thus we are living out our own personal stories that *appear* to look and feel very real to each and every one of us. We are so part and parcel to this story that the likelihood of beholding reality that is always right here, right now, is impossible. It is in this context that as so-called human beings, we are stuck with our thoughts, ideas, and concepts about reality. Words can't actually capture that which is ineffable, utterly indescribable. Again, it *seems* we are stuck with the conceptual, and not the actual, because essentially, "we" are conceptual ourselves.

If one were to speak of the human picture, it *appears* that we have come a long way from our primitive ancestors. The temporal-mind, the purported interloper that poses as your personal self *appears* — at least to itself — to evolve by separating and dividing conceptual thought forms repeatedly. Speaking on the basis of *appearances,* as children we are taught what things are from our parents to our grandparents, from a blue sky to a tree, from a cat to a dog, from the sun to the moon, and there is agreement, conditioning, education, indoctrination, all of which are reinforced and solidified by our so-called personal experiences. Our experiences solidify the knowledge, and the knowledge we accumulate is reinforced by our experiences. It's a repetitive cycle that goes on and on throughout our entire journey. We are taught whom we are, who Mommy is, who Daddy is, Grandma, Grandpa, Aunts, Uncles, and the list goes on. We are taught what objects are, and about the various things that *seem* to *appear* in our environment. Both teacher and student are inseparable. This would all be the activity of a never present temporal-mind that is constantly seeking to ensure the continuity of its pretense.

Speaking strictly on the basis of the human spectrum, it *appears* that there are separate people with separate minds. The entire human scene *appears* be one dreamlike mental construct. In reality there is no you, no me, no us, no them, as the *seeming* self that *appears* in a *seeming* human dreamlike spectrum would be the temporal-mind's self. This means that in reality there exist no "we" at all, as a purported collection of separate dreamlike thought selves has no more validity than a thought dream itself. *That which is not,* plus *that which is not,* plus *that which is not,* doesn't equal *that which is.* There's no way to start with non-existence and arrive at existence.

If one were to speak of what *appears* as the human spectrum, it *seems* as though humans are conditioned, educated, programmed, and one could even say, indoctrinated. As we *appear* to develop and age in this time-based thought construct, a personal identity, a sense of a separate self, *seems* to develop and is fortified through our experiences. As characters in this dreamlike mental construct, it *seems* we are indoctrinated to put all of our faith and confidence in what we are told is a physical world of form, that *appears* to be outside of a physical body that we have been told is our body. From the moment we are born we're taught that the physical world is real, and this passing realm is all there is, and if there is something more, we'll never truly know it until after our death has taken place, or perhaps never. Many of us are told that we must rely on *hope, faith, and belief.* This process of building up a sense of self, a sense of a personal identity separate and apart from others begins quite early, and it continues on day after day, and year after year, until it goes on autopilot, and as this separate self we begin to add to this complex structure on our own. It *seems* as though we believe in who we are, what we are, and it is confirmed again and again by our experiences, and by the society in which we *appear* to live. This pseudo identity becomes the most important factor in our experience. We use our memory to access information in order to maintain this identity, and as time *appears* to continue passing, a complex thought structure is built up representing who we think we are, what we stand for, and what we've been through, and we use this story of "me" in order to separate ourselves from others who have their stories, and to explain to ourselves the motivation behind what we're doing at different points in our lives. Essentially, we use what is ordinarily called memory to reinforce the continuity of this false identity, and it confirms the belief that we are here. This is all thought, just one *seeming* thought dream, and the only respite from this cyclic effort *seems* to be when we go to sleep, but even then, it *appears* that we have dreams, though many of us don't *seem* to remember them, nor did we decide to have them. It's as though characters in a dreamlike mental movie are experiencing yet another dreamlike movie. The divisive thinking mechanism *appears* to be continuously at work. First thing in the morning, the memory kicks in; "let me think, um, what do I have to do today, oh, that's right, I have to do this, I think I'll do that, I better get going on this thing or that thing," and the perpetual effort to maintain the continuity of this pseudo identity is engaged. This purported separate identity, this personal self is entirely reliant on the information and experiences provided by others, and if all of this wasn't available then there'd be no way for this fictional self to experience anything, and no way to maintain this fallacious identity. Not many people stop to contemplate the fact that what they *think and believe* they know about who they are, even what they are, was provided by others. All of us, each and every one of us, rely on the knowledge and information that has been handed down from generation to generation in order to communicate, have experiences, and even to ponder our own existence or lack thereof. We access what has been passed down to us from others, not only the knowledge and information, but we rely on the experiences of others as well. We interpret their experiences, listen to their stories, and over time we fabricate an illusory personal narrative, the so-called story of our lives. Information *appears* to be more readily available, and it *seems* that the hunger for more has never been greater because, as the complexity swells, then the demand for solutions is swells, only resulting in the manifestation of yet more problems that demand attention and resolution. It's a vicious cycle.

Armed with a sense of self and thus a personal point of view, we *appear* to go out into the world convinced we are separate from others, so we seek agreement and experiences that confirm and cement who and what we think we are. If we find people we agree with, at least for the most part we have a relative reality with them, generally, we agree with the way they see things, and we're able to get along. Then there are those who we're not able to fine agreement with, and so we have very limited relative reality, and eventually we have distance, lack of understanding, and conflict. It's basically automatic, and it functions on its own in a very mechanical way. All of the *seeming* content, including the characters and various facets of societies are part and parcel to the same one temporal mental dreamlike construct, the same non-present interloper.

It *seems* that for the vast majority of the characters that *appear* within this mental construct the human body is their identity, so the combination of neurochemistry, knowledge, conditioning, and perhaps even DNA, *seems* to allow people to function in the world under the assumption that they are physical entities with total freedom of action. If we were able to look at things truthfully then we would soon comprehend that given the same combination of knowledge, neurochemistry, conditioning, DNA, indoctrination, genetics, environmental influences, etc., any one of us could be that person that we so often point to and condemn. What does this actually mean? It means we are each accountable for our actions, and yet within this temporal dreamlike construct, we have little choice but to be what we *appear* to be, and it's difficult if not impossible for us to be any other way. Even if we *appear* to change on the surface, first the change was made using the pre-existing structure that was already there in place, and this change is superimposed and held in place through the use of forced effort, or discipline aimed at reconditioning. Of course, the way this *seems* to work is advantageous to the survival and continuity of the thinking mechanism itself, because when what we determine and define as bad things happen, we habitually point to those who we have predetermined are the bad ones, and then to those who we have predetermined are the good ones, and this strengthens the separate structure, and allows us to elevate ourselves over others and separate the good from the bad, the smart from the stupid, etc. This dreamlike spectrum operates in a rather deceptive way, but it is possible to sort of deconstruct, or break it down, through a process of reverse engineering. Perhaps someday computers will assist us in this area. At least for the moment, it *seems* we have little interest in how and why this world we *seem* to experience *appears* the way it does, as reported by and interpreted through the limited human senses. It *seems* that survival at various levels is what drives us, but it is this never present wannabe — this temporal-mind — that seeks survival. It is not a conscious entity, and thus, it is not even aware of what it *appears* to be doing.

When it comes to the so-called human picture, it *appears* as though the mechanism resulting from *the subtle shift* uses numerous conceptual mental constructs in order to strengthen and expand an already bloated structure. With respect to the temporal-mind, it's always about ensuring the continuity of its pretense. For example, take the simple concept of what is typically called failure. Speaking strictly on the basis of the human spectrum, it *seems* the saying, "be careful what you wish for," may be more than a casual admonition. When one embarks on a mission to do something in this spectrum and fails, it *appears* one is forever the effect of that failure. This manifest in many ways throughout societies all over the world, and naturally some are more obvious than others. Some people, for example, fail at a job and simply try another career, not the end of the world as the saying goes. But let's say for example, you're a psychiatrist and you set out with a goal to make what you determine to be an insane person sane. A dangerous game indeed, for if you fail, which is more likely than not, you go effect of that which you had set out to do, in essence, you may go a bit insane yourself. Statistics reveal the validity of what is stated here, as psychiatrist have a relatively high rate of suicide in comparison to most of the other professionals in society. It *seems* that in our eagerness to find solutions, and to initiate change, we get so wrapped up in the process that we ignore the potential consequences. This is another example that exhibits how we, as illusory separate selves, *seem* to project and reinforce our illusory reality. These purported separate selves have no independent choice or freedom of action. It's all part of one temporal dreamlike construct, a would-be spectrum that incessantly seeks to ensure the continuity of its pretense.

If one were to speak of this never present temporal mental construct, it *appears* that humanity has had a long and complex history. The momentum of thought *seems* to be unrelenting, and the numerous characters *appearing* within the context of this spectrum are forced to live in service to it, agree with it, or be crushed under its weight. It can be said that what *appears* as the human scene and its world is nothing more than vibration, and despite its growing complexity and relentless momentum, there is little more to this *apparent* relative reality than agreement. It *appears* to be an agreed upon construct that has been disrupted by a divisive mechanism that *seems* to be bent on transforming it into a nightmare. But there really is no agreement, just the one finite temporal-mind posing as many personal minds. The human picture *appears* to include separate selves with separate minds, but this is how it *seems* from the standpoint of a character *appearing* in a temporal mental construct.

The so-called physical world involves a multitude of complex factors, but like other fields of study, science is working diligently to break everything down, dividing, labeling, compartmentalizing, and interpreting, so that there is an agreed upon jargon that can assist in facilitating communication within scientific circles. Scientists have made certain observations regarding the nature of the material world, but they are relegated to observable, measurable phenomena. And there's nothing wrong with that, it's their job. Life itself is not a phenomenon, hence, cannot be studied, analyzed, or researched like conventional finite, observable, measurable forms.

In the interest of clarity, I'm going to discuss some of the basic precepts behind the scientific conclusion that this so-called physical universe is not physical or solid at all. Matter in this realm *appears* to be built on the basis of simple arithmetic, and as stated, according to scientists it's not as solid as it *appears*. The basic building block of matter is what we call the atom. An atom is made up of three essential components: protons, neutrons, electrons. Electrons orbit at a tremendous speed around the nucleus of an atom and form the entire outer shell of the atom. This whole arrangement is held together by electromagnetic force. What makes one particular type of atom different from another is the number of electrons and protons. From a human point of view, most of us know a little bit about atoms and the sub-atomic particles that make up an atom as mentioned here, electrons, protons, and neutrons, but not so common in ordinary discussions are what's called elementary or fundamental particles, such as, quarks, leptons and bosons. Science informs us that atoms consist almost entirely of empty space. In other words, the vast majority of what *appears* to us as solid is not solid at all, yet we experience it through the so-called human senses as though it is solid matter. The human senses *appear* to be designed to accept the illusion of solidity caused by the extremely rapid motion of atomic particles. When we look closer and closer at what we take for granted as the physical world, as solid, you find it's difficult to grasp anything at all. I don't want to get to technical and scientific, but the point being made here is that the closer we look at physical matter, the more it disappears. And yet we all *appear* to go along pretending that we are separate individuals living in a physical world made up of a multitude of separate physical objects. It *seems* that we have little choice if we want to have any chance of living and surviving in this world as it *appears* today. In fact, it is delusional to *think and believe* that we as individuals have a choice, or that we can somehow escape this temporal mental movie, where it *appears* to be virtually impossible for us to live in harmony.

This temporal mental construct where we *appear* to live out our personal stories has its own manner of functioning. What we consider and agree upon becomes *what is,* and if we alter *what is,* it *appears* as though we create more persistence. The mistaken assumption here from a personal perspective would be with the word is. When we say something *is,* if we looked a little closer, we would see that it really never is, so what we as humans assume to be reality is not real at all, as reality would be what really *is.* Like humanity itself, it *seems* to arise within a field of time, a movement that never stops to be, and thus, it never actually *is.* For the purposes of our discussion here, we have to look at this in the way people within the context of this temporal dreamlike construct ordinarily perceive things. There are purported teachers, guides, and gurus that say emphatically that one must accept *what is,* even though they *appear* to be pointing to what never *is,* or *that which is not.* So, it *appears* as though we have a case of mistaken identity. The word *is* means, "to exist, be present." Because there is no way to capture the present in time, nothing within a purported field of time ever really *is*. However, in a personal context when we say something *is,* we are referring to our own personal interpretation of what we *think and believe* we see. And although we *seem* to have a false assumption regarding reality or *what is,* it *seems* our ideas about reality or *what is,* have evolved right along with the never present temporal realm, and *what is* — what we believe to be reality — *seems* to have its own manner of functioning. In other words, it can be observed and deconstructed. For example, within the context of this temporal mental movie, if one sees something exactly for what it is then it cannot survive, at least for the viewer it *seems* to vanish. Everyone has had the experience of attempting to remember something they wanted to tell someone else, and the moment it *appears,* in an instant it vanishes. It's already there in the archives, in the database. It exists as a thought construct, a concept, or a mental image, you search the database, you have an idea of what you're looking for, but you can't remember exactly what it is, then suddenly there are two, and they appear to cancel each other out. You almost had it, but in an instant, it's gone! After some time passes and you are totally immersed in something totally different, bingo, it *seems* to *appear* out of nowhere.

From a human or personal standpoint, the way something *appears* is defined as the way it really *is*. In this dreamlike construct when we personally try to ignore, reject, or resist the way *think and believe* something *is,* then we strengthen what we *think and believe* actually *is*. This dreamlike mental construct is entirely dependent on agreement, and therefore, *that which is* has its roots in agreement, and by rejecting and resisting it, we are actually fortifying an illusory agreement, and consequently it *seems* to become even more real in our experience. I don't want to go too deep into this subject, as it can be a bit perplexing, but it's fascinating stuff in terms of how the mechanism works to perpetuate its illusory status. Certainly, it can *appear* to be somewhat confusing because personal sense sees *that which is not,* and believes it to be *that which is,* and this is yet another reason why human sense experiences this dreamlike realm as reality. The bottom line is that upon closer inspection, this purported spectrum that we personally *appear* to live and function in on a daily basis is rather flimsy, and the "we" spoken of here would be inseparable from this spectrum. In effect, the characters are products of this dreamlike mental construct, indeed, inseparable from it, and thus, they *seem* to testify to its purported reality.

Assessing things on the basis of what *appears* as the human scene, most humans experience the process of living as though they are separate not only from each other, but from the world that *seems* to be around them. Conversely, quantum physics suggest that everything is interconnected and made of the same basic stuff called energy, vibrating at varying frequencies. This *seems* to imply there are separate things that are connected with each other. Words tend to be insufficient when we start discussing what we call life. Whatever is said is not it. The best one can do in this scenario is to use words as clumsy pointers to point to that which can't really be pointed to or at. From a perspective based on human or personal sense, life *appears* to be a multitude of *seemingly* separate things that are actually interconnected because this temporal mental construct is vibration that *appears* to be dualistic in nature, and from a viewpoint *seemingly* within it, one is stuck starting with a divided perspective. It's never about the words, as the word life is not life itself, and anything said about life is not really it, but it *appears* these words are the best tools we have within the limited context where words *seem* to have a valuable and functional role. There are many synonyms for the word life, and though none of them can really meet the occasion when it comes to defining or describing *that which is all,* some words *appear* to work better than others when it comes to discussing that which is essentially wordless. The human picture *appears* as a realm of forms, limitations, opposites, a world that is measurable and observable. Starting from an illusory point within what *appears* as the human picture, it's not possible to capture, much less express, that which is formless, unlimited, non-dual, immeasurable and invisible to the human senses.

Just imagine being inseparable from the totality of life, the totality that life is, literally! Personally, you as you know and experience yourself wouldn't even *appear* to be present. The only one present would be the present itself, the all-ness that is life being *all there is.* It *appears* that many of us have sensed that there is a reality or actuality that *seems* to be underneath what is more or less a dreamlike spectrum where human experience *seems* to play out. Words simply fail when it comes to this subject. Whether called life, reality, or the infinite, what is utterly ineffable, indescribable, and undefinable, is not a vast physical phenomenon that can be observed and measured. It's actually not any type of phenomena. It can't be reduced to a word concept. Even to use the word "it" makes no sense, and words like is, reality, truth, awareness, the infinite, *seem* to carry way too much baggage. What we all *seem* to take for granted, what *appears* to us personally as reality, is not necessarily the way things actually are, not reality itself, and all of it may very well be an illusion, but in order to survive as characters in this *seeming* dualistic world of limitation, it *seems* we must play the game the best we can, or risk being ejected from the game. Looking at the human spectrum as it *appears* today, it turns out the way things are progressing we may all be expelled simultaneously. It *seems* as though we have reached a pivot point, and certain fundamental issues require our attention in order to ensure the survival of the human species. We can sit around and theorize what human thinking is and what it isn't, but then we are left to live like hamsters on a wheel simply thinking about thinking. This evolves into a vicious cycle where even *apparent* solutions only *seem* to feed the dialectical nature of thought. There is another possibility, perhaps a solution will arise on its own from a point that is not rooted in never present human thinking. Maybe only what we call nature can offer an antidote that will slow the momentum of this mechanism that may be the demise of the human species, yet nature itself has not been spared from *the subtle shift,* that being said, what *appears* as a threat to humanity, like a virus, or a climate crisis, may also be what acts as a catalyst that puts humanity on a more sustainable course.

The main focus of this book, i.e., *the subtle shift,* may sound serious, and no one is denying what *appears* as the human picture, but things are not as they *appear*. It's never about blindly believing what some author says, so take a look for yourself, dig at what *appears* as the human picture and discern for yourself what is fact. If it comes from another then this leaves one as a believer who lives in bondage to a concept. Following another's path puts you on their path and not on your own, but you may discover that there actually is no path, as that which *is,* already *is,* meaning there is no necessity for a path, as now, the present, is the only self that *is,* and it is already at itself. Being *all there is,* there can be no right approach or wrong approach to reality. In truth, there can be no approach to reality, and no one to approach it.

The human picture *appears* within the framework of time, but time would be when it is not now. When is that? Never! Now is not a time, not in time, nor is now a product of time. That's reality, the way life is right here, now. In one's daily experience of living everything one encounters is fleeting, passing, changing, and ultimately leaving. Surely energy or vibration that is constantly changing, shifting, ebbing and flowing, can't be reality, because it's never really being, never actually present. Only reality or being itself is being. This is not a book on reality, and although discussed a bit here, it's important to note that it's never about the words. The reason the subject of reality is mentioned here is because discussing reality *seems* to highlight the ephemeral nature of what *appears* as the entire *seeming* temporal human construct and its utter lack of reality. Indeed, it would be lack itself, as it really has no presence to speak of. It is not being, not present, not now, hence, not, period!

CHAPTER V

TIME

Time *seems* to be a mysterious topic. What time was it when time purportedly started? What time was it when someone started keeping track of time? In other words, how much time had already passed before time purportedly began? When did now, the present, vacate its own eternal present-ness, allowing for never present time to begin? In the absence of human ideas about time, is there any such thing as time at all? Humans *appear* to move along with the flow of time focused on a future *that is not yet,* that instantly becomes a past *that is not anymore.* Time *appears* to flow non-stop, but it never stops to actually be. Time is not being, not present, not now, and thus, time would literally be a state of not. And it's really not possible to be a state of not, as it is not, it doesn't exist. This means there really is no such thing as time, and consequently, it means that the entirety of humanity *appearing* within a passing dreamlike framework called time, is not. Surely this will sound crazy to human thinking that is synonymous with time itself, but there's no rational reason why one would rely on *that which is not* to inform one as to the nature of life, or *that which is*. As stated, time never stops to simply be, it's always moving on, passing away, and therefore, anything *appearing* within the context of time is relegated to not being present, not being, literally. This means there can be no such individual entity called a human being. With the exception of brief "glimpses," fulfillment from a personal or human perspective within the framework of time that is never present, is always going to arrive tomorrow, or at some point in a distant future, never now. All wanting and seeking involve becoming, and thus, all involve time. Time would be when it is not now, and because it is *always* now, time is not. Time is literally never now, and thus it is inseparable from lack and limitation. This means one is total, whole, complete, fulfilled now, or never will be. And now is not a time, now is time-less, the utter absence of time. Not even now has been now before. Now is ever present, ever fresh, ever new. One can't escape the now, nor can one make now not be now. You're welcome to try but you will undoubtedly fail.

If one were to speak on the basis of *appearances,* it's not possible to discuss time without discussing space. If you see anything change position in space then it *seems* as though there had to be time for that to occur, but time can never stop to be, so it's an illusion. Time and space are both products of thought. If thought isn't there, then you as you know and *appear* to experience yourself personally are not there. In this scenario there would be no recognition, no separation, and therefore, there'd be no space, no time, and no pseudo separate or lesser self. This temporal dreamlike construct we *appear* to live in is not only a product of thinking, it's synonymous with thinking, and its main mission is to ensure its continuance without a moment of stoppage, and thus, we *appear* to experience a world where the only constant is change. And this "we" spoken of here is part and parcel to this changing relative spectrum. Even from an illusory personal perspective, one has to sort of create time or it *seems* one will run out of it. This would be the activity of the temporal-mind, as it *seems* to keep dreaming or thinking up the human picture moment by moment. It's not reality, not being, but to the never present temporal-mind, it *thinks and believes* that it is real. Thinking is a process that involves change, hence is never being, while being itself is not a process at all, as it is changelessly being.

Because you as you know and experience yourself are not separate from the thinking mechanism that resulted from *the subtle shift,* it only makes sense that if you desire change you will persist, and if you refuse or resist change you "personally" will not persist, you will be crushed under the weight and momentum of a conditional mental construct that has as its mantra, change, change, change, change, etc. You as you know and experience yourself create time by initiating change. This "you" or "self" would be the temporal-mind's manufactured self, pretending to be your actual identity.

When speaking on the basis of what *appears* as this temporal dreamlike spectrum, what we *think and believe appears* as the world we experience, and the "we" spoken of here is a product of the dream itself, and inseparable from it. If some "thing" *seems* to *appear* within one's experience, and one alters it, then one creates persistence, and if one has persistence one has time. It is this illusory persistence alone that makes, *that which is not, appear* to be *that which is,* it makes it *appear* to be. This *seems* to be the case with every aspect of how we personally experience our world. Even a lie will *appear* to persist because it has been altered.

Speaking on the basis of what *appears* as our daily experience, it's not at all surprising that pleasurable experiences *seem* to be fleeting, whereas not so pleasurable experiences *seem* to go on and on as though they have a life of their own. Upon observation, it's simple to explain this phenomenon. One does not try to alter an experience that one has deemed pleasurable. When it comes to those experiences that one interprets as not so pleasurable, there's an impulse to change them, resist them, alter them, fix them, correct them, etc. There's a demand for change, a demand for things to be different than the way they *appear* to be, and as a result we often get a persistence of that which we didn't want. One can see how the divisive mechanism that resulted from *the subtle shift* works in various facets of one's daily experience. Time and space are inseparable from thought, so if the thinking mechanism seeks to continue it makes sense that one — thought — would seek to alter what one *thinks and believes is,* or change what one *thinks and believes is,* in order to create time and an illusory persistence *that which is-not.* This becomes acutely obvious when you tell someone that they're dying, and they only have a short time to live. Before you know it, they get very busy. This has to be handled, that has to be done, better take care of that, change my will, start moving some things around in the house, get rid of stuff, have to get my hands on one of those so that other thing can be resolved; must get my affairs in order is usually the basic explanation. And the days, months, and years go by, and there they are, alive! They got busy changing things and changing the position of things in space which resulted in the *appearance* of more time. This thing called time *appears* to be a bit mysterious when it comes to how it functions, but to the divisive mechanism that seeks to ensure its continuity, change is essential. Thought can't stop to be, so it must continuously change. This is why change *appears* to be the only constant in this *seeming* temporal mental construct. Time and change are inseparable, akin to an illusory coin. One side would *appear* to be time, while the other would *appear* to be change. It's easy to see that it's not possible to have one without the other. In truth, the entire coin is non-existent, as it has no basis in reality — that which truly *is*.

This dreamlike mental construct that we *appear* to live in has its roots in a lie. Even a pleasant dream can be considered a lie, in the sense that it's not real, not truth. The point being made here is that if you alter a lie it *seems* to result in a greater persistence. And this *appearing* to be the case, it *seems* there are these ancient thought structures that many of us believe in wholeheartedly simply because they have been around for a protracted period of time. It *appears* as though there's an unusual reverence and respect for anything that has purportedly survived down through the ages. It *seems* that there are not many amongst us that really look into this particular phenomenon, but the fact remains that things persist as a result of change — as a consequence of being altered. It has a similarity to what happens when one tells a story that gets passed along from one person to another, by the time you go down the road a bit the story is hardly recognizable when compared against the original. The act of altering the story at every step is what *seems* to empower its illusory persistence in this temporal mental construct *appearing* as the human picture.

The main focus of this book is on a profound shift in vibration, i.e., *the subtle shift,* that *seemingly* altered the entirety of what *appears* as the human picture. For the purposes of what is being discussed here, the emphasis is on the central components of past, present, and future, as this is how time is defined within the human spectrum itself. Even having a discussion regarding the now, or what is sometimes called reality or the present, can, *seem* somewhat silly as we are discussing it from a position within the context of this temporal mental construct that's in a constant state of change. Also, it takes time to say anything about now, so as mentioned earlier, in reality nothing can really be said without *seemingly* using the very thing that this now that is now precludes. In this context it *appears* we are stuck with a paradox, but only if one *seems* to ignore now, one's own present-ness. Starting with, indeed as now or the present, only the present is present, hence, there is no paradox to now itself.

If one were to speak on the basis of human sense, in order to experience anything, we *seem* to be totally reliant on what we ordinarily call our memory, so everything is interpreted and filtered through the past. In reality, there is no time in the present and no presence in time. This is the basic dilemma facing humanity, as personally, one's world and one's entire experience of said world, *appears* within the framework of passing time. The word *time* means, "the indefinite continued progress of existence and events in the past, present and future regarded as a whole." Obviously, the human definition of time is fallacious as time can never be present, yet we *seem* to go merrily along ignoring this false assumption.

In many so-called spiritual circles, it *seems* there's a lot of talk regarding how to live more in the now, or how to be more present, but can we personally say anything about the past, present, or future without relying on the past, without using our so-called memory? One can try but it's really not possible. Starting under a false assumption, as a separate self with a separate mind, functioning within the context of time, it would be thought attempting to live in the now or be more present, and it always fails. First and foremost, time is never being, it's always moving, passing on, and secondly, it's utterly impossible to capture now, or the present, within the context of time. This means anyone trying to live in the present, or now, is working in futility and will never get there. Only now is now, and now isn't busy trying to become more now than now already is. The present is already present, and it's not busy striving to become the present it already is being presently.

Time — *that which is not* — and all that *seems* to arise within the context of time is not present, not being, not now, and thus, not. Despite how things *appear* to the purported human senses, *that which is not testifying to that which is not, in a place that is not, at a time that is not, is not, literally!* One cannot be what is not now, as what is not now is not being, and therefore, nonexistent. The word *nonexistent* means, "not existing, not real, or present." Certainly, one cannot exist as nonexistence. It's not possible to be not real. One cannot be what is not being. To reiterate, the word *be* means, "exist or be present," and the word *being* means, "existence." This means that if one is not being one does not exist; one is not alive. It's actually not possible to be *that which is not,* as *that which is not,* is not, it has no existence at all. Existence does not co-exist with a measure of its own absence.

It does *seem* one can say a lot of lofty sounding things about now. For example, one could say that now is not a brief moment between a past and a future. One can state that this now that is now is timeless, formless, changeless, boundless, but then one is using time to make these statements, and of course, it must be noted that words and thoughts about now are not now itself. They are sort of one step removed, and it must be noted that one *appears* to be thinking about now through the lens of the past. In truth, nothing can be said about now, for whatever now is precludes words and thought altogether. Even that statement is not entirely accurate, as in truth there's nothing to preclude. What does this mean? It means any and all seeking and searching has its origins in a completely false assumption. This assumption suggests one can start as a separate personal or secondary self, and become, arrive at, enter, or merge with now, the only self. This is simply not possible, as both assumption and assumer, belief and believer, dream and dreamer, have no foundation in reality. It is yet another trick, another method thought employs to ensure its illusory continuity. It is by initiating this futile process of seeking and thinking about something called now, that thought ensures the continuity of its pretense, at least to its never present self. By no means am I denying that reality is now, as reality is now, literally, or not at all. In fact, there is no power of now, just power that is now, literally! Word concepts can't do the job when it comes to infinite reality, now, presence, or whatever word one wishes to use to clumsily point to *that which is all there is of all there is*. The point being made here is that from a personal point of view it's all speculation, and if it exists as existence itself, it's not something that can be experienced like other ordinary finite experiences. As soon as you said you did, it's very clear that you didn't, and you're left interpreting yet another experience through the use of and within the context of never present thinking. Certainly, it *seems* one can use thinking and memory to expound on now, and how it is always now, etc., but then all of this would seem to require time that is not now. Even to use the word "always" implies now is now over a period of time. This is utterly false! Now precludes time, because time — thinking — can never be now.

As human beings that are never actually being, we must come to terms with the fact that we *appear* to be part and parcel to a sort of temporal mental movie. The "we" spoken of here will never break free, as this "we" is inseparable from the temporal mental movie itself. The various human ideas and theories in regard to now are speculative conceptual constructs made within the context of time, and this thing we call time *appears* to itself to be a movement, a continuity of what is called the past. Even a so-called past or future *appears* to be a thought that one *seems* to project right now because *only now is* — it really can't be any other way.

From a personal perspective, as we *appear* to move about our day, it's impossible to be present. If you took a pin and tried to stop the hands of a watch in order to capture this moment, the present, you would always just miss. Speaking on the basis of what *appears* as the human picture, as soon as one interprets the present it is incorporated into the past, and the same goes for the future. Starting from a point of view that *appears* to be rooted in time one can't really interpret the present at all, as the present can't be found much less captured within the framework of time. Time doesn't actually exist, and the present is existence itself. The so-called future *appears* to be shaped through the lens of the past. The thinking mind even attempts to interpret now or reality through the filter of the past. It *seems* as though starting from a finite perspective in time, personal sense seeks to place reality, now, into a frame so it can reduce it down to a concept — an idea that has its roots in a *seeming* mental dreamlike construct that is never now. Starting with the only valid premise there is — *only now is* — nothing can really be said about now, there's no division in now, no prior to now, no after now, and nothing to contrast against this now that is now. Whereas the observable world of form *appears* to be littered with dualities, this now that is now has no opposite. It is simply now. Now is *that which is. That which is not* is not the opposite of *that which is* because it's not here now, not present, to be the opposite of anything. While it is stated that nothing can really be said about now, it doesn't *seem* to stop some that *appear* amongst us from doing just that, and this is just fine if one understands that it is "never" about the words but about what *seems* to underly or be behind the words. Whatever that is, it is *all there is,* there is only one *is,* one now, one all, one infinity. Because this *is,* is *all there is,* it's always about what it is to itself. The way life *is,* is that it is the one that actually *is.*

Starting — abiding, beholding — is now, and one starts with this premise because it is already so, already the way life is. Starting here, now, there is no personal you, no little self, just now being the only you, the only self that *is*. So, any discussion between purported separate individuals regarding now is futile and just more dialectical thinking. In order to speak of what is real, true, valid, one must start with that which actually *is,* indeed as reality, and in this context, the only valid context, it's never about the words, thoughts or concepts, but this now that is now, this alive, ever-present, all-inclusive reality that is wordless, indescribable, ineffable, yet right here, now.

Time is not just a product of thought, time and thought are inseparable, both are dependent on and correlate to change. So-called human thinking uses its own predefined concept of time in order to create the illusion of separation between a past, a present, and a future. It *seems* that the thinking mechanism is only concerned with its continuity, and if it *appears* that there is time, then there is continuity, otherwise there'd be a complete absence of a *seeming* linear mental construct, and there would be no continuity, no time. The word *continuity* means, "the unbroken and consistent existence of something over a period of time." It is the *seeming* continuity of thinking that *appears* to produce the never present illusion of a temporal dreamlike mental construct filled with separate characters, separate selves, with separate minds. Only existence exists, so the definition of the word continuity is entirely fallacious. Essentially, it *seems* as though time imagines that there's a period when the present is absent — when now is not. This is simply impossible. If you disagree then by all means, try to make it not now, try to be aware 10 minutes ago, or 10 minutes ahead, try to make the present not be present. Good luck!

CHAPTER VI

DUALITIES

At its core the temporal-mind *appears* to be energy that is constantly moving, reacting, vibrating at varying frequencies. Vibration is duality, so this temporal mental movie *seems* to *appear* as a world of opposites, the past, *that which is not anymore,* or the future, *that which is not yet.* The temporal or time-mind can't deal with now because now has no opposite, and now is not a time. Now is all, the alone one, the sole self, hence no vibration is possible.

Given that the temporal-mind's thinking or dreaming is nothing more than vibration, it *seems* only sensible that one would *appear* to find countless dualistic relationships that *appear* to touch on every facet of the *seeming* human experience. Within the context of this temporal mental construct, each dualistic relationship *seems* to be composed of two inseparable aspects resting on the same spectrum. These dualities are so interwoven, not only with each other, but also throughout the entire purported human picture, that it's impossible for one to imagine the full magnitude and degree of complexity involved.

If one were to speak on the basis of human or personal sense, it *seems* that the divisive mechanism that resulted from *the subtle shift* dominates virtually every aspect of daily human experience, so it's impossible to talk about I, you, me, them, us or we, without at the same time acknowledging the *apparent* ongoing conflict between the divisive thinking mechanism, and the organism that is ordinarily called the human body. In respect to this book, words like I, you, me, them, us, or we, are used strictly as grammatical conveniences, because it would be the temporal-mind itself that not only generates or dreams up but experiences the illusion of separate selves with separate minds. The divisive thinking mechanism resulting from *the subtle shift seemingly* has transformed thought from that of a servant to master of the house. This mechanism is a mutated version of what is typically called human thinking. Yes, it *appears* as though there was a *subtle shift,* a shift in vibration, a change in how the temporal-mind thinks or dreams, and it *seems* to be reflected in what *appears* as the human spectrum. Questions about when, why, or how this shift or reaction in vibration occurred *appear* to be meaningless. Searching for the origins of that which is never present would be insane. So, it *seems* we discuss this so-called mind, this absent interloper if you will, without any fear or trepidation, with certainty that it has no foundation in reality.

While one may *appear* to be functioning with the clarity that the temporal-mind is literally not present, not being, not now, and thus not, period, this doesn't mean one disregards or doesn't care about what *seems* to *appear* within the context of the human picture. It just means one *seems* to have an entirely fresh perspective. One isn't tricked or fooled by *appearances*. In effect, one knows that the way things *appear* is not the way they are. The reader reading these words is not merely one of many characters in a temporal mental movie, but in a personal context this may *seem* to be the case. The profound shift in vibration that caused a genetic mutation triggering a neurological disorder called *the subtle shift, appears* only within the context of a temporal mental dreamlike construct. The temporal-mind responsible for this construct has no foundation in reality, as it is vibration that is never present. No "body" can tell another as to the nature of reality because reality precludes otherness entirely. Don't trust some author, ask yourself and be honest. Can reality be a never present vibration? Does it make sense that reality would ebb and flow, come and go, arise and recede? Wouldn't it make more sense that reality simply *is?* Would reality have a frequency? Does it come in and out? The answers to these questions are self-evident, but only to reality, the only "self" there *is*. This means that the entire human picture, inclusive of its many forms, whether called physical forms or thought forms, essentially its entire universe, has no existence at all, as it never really *is*. Thinking, sensing, imagining, emoting are all part and parcel to this one, never present dreamlike mental mechanism. It thinks, dreams, or projects its world, and it is the one *seemingly* experiencing it.

If one were to assess things on the basis of what *appears* as day-to-day human experience, it *seems* as though thought has elevated itself, and assigned itself a godly status in its own illusory thought universe. This divisive mechanism is not only deeply rooted in the structure of the purported individual, or the personal self, but *appearing* as a duality, it is its own world as well. So, there's not a multitude of separate entities with separate minds each dreaming and experiencing a unique dream. There's just one *seeming* dreamlike mental construct that is-not, one fleeting, shifting, ebbing and flowing spectrum that is not being, not present, not here, now. As individual characters that *appear* within a dreamlike realm, it *seems* we have little choice but to accept this *apparent* reality as it is imposed on us or be crushed under the immense weight and momentum of a mechanism that *appears* to have no off switch. What we are irrespective of words, ideas, concepts, mentation, and interpretations, we personally *seem* to have no way of knowing. We can only speculate, pontificate, articulate, and inquire. It's a futile game that ultimately ceases when *our* illusory personal story ends. Then what never really began recedes into that from where it *seems* it sprung forth, oblivion.

In this book, twelve dualistic constructs have been identified that directly correlate to issues that *appear* to have beleaguered humankind for centuries. Speaking on the basis of what *appears* as the human spectrum, it *seems* as though these issues can be traced back to humankind's inequitable handling of these particular dualities. It's not necessary to delve too deeply into each individual duality. The divisive mechanism that seeks to be master of the house *appears* to approach each duality in exactly the same mechanical way. The twelve dualities are as follows: *birth and death, happiness and unhappiness, health and sickness, wealth and poverty, love and hate, security and vulnerability, peace and war, success and failure, compassion and cruelty, pleasure and pain, hope and hopelessness, freedom and imprisonment.*

Dualistic relationships *seem* to be the basic building blocks of what *appears* as one's day-to-day human experience. Because humanity and dream are one in the same, not one of these dualistic relationships can be considered without first *assuming* the fundamental duality upon which all the others *seem* to be built. This primary duality is often *assumed* to be *life and death*. Upon deeper inquiry, this duality that is ordinarily viewed as *life and death* is actually *birth and death*. In the absence of this primary duality none of the others have any status at all, not even in a *seeming* sense. Life is not an aspect of a dualistic relationship. These dualistic constructs *appear* to be quite real, at least to the temporal-mind itself. This entire mental movie *appears* as a duality, and therefore, all the *apparent* dualities within this dreamlike vibrational spectrum are part and parcel to what is *apparent,* but they have no status in reality, what is actual, or the way life is, right here, now. Life is entirely non-dual, non-vibrational, yet utterly alive. Life doesn't co-exist with another. There is but one life, one now, one presence, one all, one non-numerical infinite reality. The author is not suggesting the reader reading these words blindly believe in what is being said here, or simply take the author's word for it. The author suggests that the reader inquire and dig at the truth of these statements and behold whether they are true and make perfect sense or not. Life or reality is incredibly simple and practical, yet being *all that is,* it can only be known or beheld by itself. Ask yourself if it makes sense or not. Would it make any sense that reality can be known by unreality? Does it sound sensible to say *that which is not* can know or behold *that which is?* Can the so-called finite co-exist with the infinite, existence itself? Given what the reader actually is, indeed, that the reader truly *is,* the answers to these questions are totally self-evident. Again, words are inherently dualistic containers, inextricably tied to thoughts, inseparable from thought, and thus there is no way to capture *that which is all there is.* Even the word "is," is too much baggage, and to use word concepts as pointers is pointless, but within the context where it *appears* this book is being presented it *seems* these are the best tools available at the moment.

If one were to speak on the basis of what *appears* as the human dimensional temporal realm, it can be concluded that humans have placed a heavy emphasis on the birth aspect of the dualistic construct designated as *birth and death.* As a result, it *appears* that humankind has generated a lot of disharmony on this planet in various societies around the globe. Speaking in terms of this temporal mental construct, it *seems* that humankind can't resist interfering with nature, even though the purported human organism is not really separate from it. Part of the confusion would have to be placed at the doorstep of how humankind defines the word nature. The word *nature* means, "the phenomena of the physical world collectively, including plants, animals, the landscape, and other features and products of the earth, as opposed to humans or human creations." Through the use of thought, it *appears* as though humankind has managed to separate itself from the totality of what we call nature. So, if one were to believe the human perspective on things one would have to conclude that humankind is unnatural, an interloper. Of course, what is typically called nature is not separate and apart from the temporal-dream spoken of in this book. Anything *appearing* in nature *appears* in time, and would be the same never present energy, reacting with its own vibrational frequency.

The incorrect assumption where this primary duality, *birth and death* was falsely thought and believed to be *life and death,* has had numerous consequences. Again, this *appears* to be the work of an interloper, human thinking! Somewhere along the line, as a result of this confusion, humans began to associate and interpret birth as new life, and in certain societies it *seems* youth has an elevated status, as it's considered newer life, as opposed to the elderly that are considered older life. In terms of American society, it *seems* as one *appears* to grow older one becomes less valuable until one is virtually disposable. Regarding the other aspect of this duality, death, it *seems* that over countless years humans have distanced themselves from death, and the entire process is often somewhat clinical with many people dying in hospitals or hospices, as opposed to at home surrounded by loved ones. What we recognize and label as life, we seek to protect, extend, and when it comes to a birth, what is often believed to be new life, we embrace, elevate, and we give special attention to this purported new life. And of course, in this context, humankind values what it deems to be human life more than any other species on the planet. What are the consequences of this approach? Humans kill and are prolific deliverers of death, not only to other humans, but also to numerous species across the globe. By over emphasizing life — as defined by humans — and the sanctity of life, humankind is the steady sponsor of death. The massive increase in human populations across the globe is already putting immense pressure on the resources necessary to sustain life — at least as humankind knows and defines it. If this trend continues, then this inequitable approach is sure to lead to an increase in conflicts, as various nations fight for a share of the remaining finite resources in order to sustain their societies. Common sense dictates that more war equates to more death. Human impact has already thrown the natural balance of numerous species *appearing* as forms of life totally out of balance. Humanity consistently treats other species as second-class citizens and unworthy, or at the very least, not a valuable as the human species, as a result many species have been forced into an early extinction and countless others are under intense pressure. This is the pivot point in regard to how humanity has approached and misunderstood this basic duality. Our one-sided approach to this duality has resulted in our getting more of that which we didn't want. Because we see life and death as opposites, we seek to distance ourselves from death, avoid seeing it, and talking about it, unless it's in the movies where we *think and believe* it's not real. In this temporal dreamlike construct that *seems* to be nothing more than a mental movie, the purported characters that *appear* within the movie believe it to be reality, and so they act and react accordingly. Within this mental construct, people are akin to puppets or automatons having no freedom of choice. Undoubtedly, it *appears* as though we have built a strange box for ourselves. This temporal dreamlike construct that *appears* as the human picture and its world, *seems* to be progressively mutating into a nightmare for many of its *seeming* inhabitants, and as conditioned automatons, it *appears* as though there's nothing anyone can do about it.

If one were to speak on the basis of what *appears* within the context of time, every second of every day there are approximately four births and two deaths in the world. It's not difficult to see the impossibility of sustaining this kind of growth in populations on this planet, not to mention the immense pressure it puts on resources. This approach translates into an increase in births for the human population, but it has simultaneously translated into death, and in many cases premature extinction for numerous other sentient creatures on this planet. It *seems* clear that thinking has allowed us to value ourselves far above all the other species on this planet, and consequently it *seems* there are many amongst us who believe these other species to be expendable. It *appears* as though we have no way of coming to terms with the ephemeral nature of this temporal dreamlike construct. The divisive mechanism that resulted from *the subtle shift* has taken a foothold as master of the house, and it doesn't *appear* that it's going to release its grip. Understanding the nature of duality and the many dualistic constructs that *appear* within this temporal mental movie is a small step, but it *appears* that there is no panacea for the ills of humankind. Within this temporal dreamlike construct, it *seems* that we are powerless to change our fate. It *appears* that everything we do is strengthening and increasing the momentum of the mechanism that is at the heart of that which is driving humanity toward extinction. Whether it is human or other species, it *seems* the basic mantra is the same, "we're here to help you but first we have to kill you."

Generally speaking, humans *think and believe* that they create life, save life, and take life, but it never seems to dawn on them that they're not dealing with life, the actual, but only life the conceptual. It *appears* that humans believe that forms have life in them, and then they proceed to project their stories onto these forms. Being infinite, life itself is never inside any "thing," any "form," or any "body." The body *appears* as a limited or finite form, and science tells us that even the body is not a solid object. The entirety of what *appears* as the human picture and its universe is energy vibrating at different frequencies. This includes the so-called human body, and essentially, this is the case for all forms. It's all vibration, moving or vibrating at different frequencies. This means the entire human spectrum and its world is never present, never being, never now. Life is not vibration. Life is not frequent. Life doesn't ebb and flow or come and go. Life is! If one were to speak of the way life is, one would simply say, the way life is, is that life *is,* literally. Life is another word for existence. Nothing caused existence, as it would have had to exist in order to be a cause. Life or existence simply *is!*

As already noted, speaking on the basis of the human spectrum, *birth and death, appears* to be the basic duality. One either *appears* to be here in this temporal dreamlike mental construct or not. In the context of this temporal mental movie, you as you know and experience yourself are a product of thought (dream). What you are absent the knowledge you *seem* to have about yourself is something that you'll never know, at least not in a conceptual context. In the absence of birth there is no opportunity for a personal story of you to evolve and take root, but once that story is *seemingly* established then numerous dualistic relationships *appear* to arise within the context of what *appears* as the human picture. This *apparent* birth is a false assumption, therefore, all that *appears* to come forth stemming from this assumption would rest on a somewhat shaky foundation, as the conclusion must contain the premise, and if the premise is a false assumption, then the conclusion has to be a false conclusion. The word assumption is synonymous with the word premise. So, what we *appear* to be dealing with, in the context of the human experience, is a fabricated or untrue premise. This means the purported temporal mental construct inclusive of the many dualistic constructs within it, has no foundation in reality. While there really is no temporal mental construct, it *appears* as though there is, but only from the standpoint of a so-called separate personal self within a purported temporal mental construct. It testifies to its own imaginary presence. The purported temporal construct *appears* as a duality inclusive of numerous dualistic constructs. Given the ephemeral nature of daily experience, it only makes sense that if at all possible, we attempt to understand them. Dualities are complementary relationships that *appear* to rest on the same spectrum. In other words, one aspect of a duality enables us to define and recognize the other. It's rather simple to see that hot and cold are literally varying degrees of the same thing. There's an old saying that states, "there are two sides to a coin but there is only one coin." If one *seems* to pick up one side of the coin, one has automatically picked up the other side. Both sides would be part and parcel to this never present dreamlike mental spectrum. If you insist on only recognizing one side of the coin as a way to avoid the other side, then it is the other side that will *seem* to dominate your day-to-day experience. Because each aspect of a duality gives standing and meaning to the other, there is a built-in reciprocity. Happiness is a great example, because so often the higher the peak the deeper the trough. Again, the demand for an unrealistic level of happiness often leads to unforeseen suffering.

If one were to gauge things on the basis of what *appears* as the human scene, the divisive mechanism that resulted from *the subtle shift* is so complex and has such inconceivable momentum, that it *seems* it would be impossible to resolve this inequitable approach to any of these individual dualities through one's own volition. What better way to guarantee its continuity than to motivate illusory separate selves to seek in vain through an approach that feeds and strengthens the mechanism itself? The temporal-mind is not a conscious entity, so this entire scenario *appears* to be the result of a continuous reaction of energy. In other words, no one planned it to work this way. No "body" is in charge. It's not necessary to go into too much detail regarding how humankind's inequitable approach to each of these dualities *appears* to manifest within the context of this temporal mental construct. The thinking mechanism resulting from *the subtle shift* is entirely mechanical, so its functionality is identical regardless of the duality in question. Whether it is: *birth and death, happiness and unhappiness, health and sickness, wealth and poverty, love and hate, security and vulnerability, peace and war, success and failure, compassion and cruelty, pleasure and pain, hope and hopelessness, freedom and imprisonment;* the way the mechanism *appears* to function is identical in each case. Speaking on the basis of what *appears* as the human spectrum, societies *seem* to assign an extraordinary value on what they have predefined as the preferable facet in each dualistic construct. The mechanism *seems* to operate exactly the same irrespective of the dualistic construct or society in question, *apparently* cultural and geographical distinctions make no difference. The consequences of our inequitable approach to these dualities are not always acutely obvious, but in most cases they're rather easy to perceive. In one's daily experience of living, in a human context, these manifestations are part and parcel to the human picture itself. It's not that difficult to see how a quest for happiness can make one unhappy, how a demand for peace often results in war, or how relying on hope can make one feel hopeless. The so-called human spectrum is littered with examples.

When speaking of these dualities that *seem* to *appear* within the context of this temporal mental construct, it's as though one is often driven by one's fear of the not so preferable aspect of whatever dualistic construct is in question. One seeks to avoid the not so desirable aspect, and *appears* to be striving toward the desirable one, but one's attention — thought — is on the aspect that one would rather not have to experience. It has to *appear* this way, because the "one" mentioned here is really inseparable from thought. All wanting *appears* to be driven by a sense of lack, and since this temporal mental dreamlike construct is synonymous with non-being, non-presence, unreality; it is synonymous with lack. If thought that never really *is,* is itself lack, then all that *appears* to arise within this temporal mental construct would be inseparable from lack. It's impossible for lack to be, as lack is not here, not present, not around to be at all. This temporal mental movie is vibration, a pattern of energy; always passing on, ebbing and flowing, coming and going, arising and receding, never stopping to be present. This would be the very epitome of lack, as it lacks presence, and presence is being *all that is being.*

Generally speaking, most wants are rooted in what one doesn't want. While it *appears* as though one is driving to fulfill a want, more often than not this drive is motivated by what one doesn't want. This is the nature of the divisive mechanism that has hijacked human thinking. If one were assessing things based on the human picture, it *appears* that because of this interloper, this machine-like thinker of thoughts, the *seeming* approach to these dualities has been predominately the same for hundreds of years. If what *appears* as day-to-day human experience is analogous to a dream then it's easy to see that there's no way a character in a dream can escape, but perhaps the dream need not become a nightmare. What you as you know and experience yourself do, want to do, and hope to do, *appears* to be initiated by an illusory self, and for the benefit of an illusory self. This personal separate self *appears* as a wannabe, a mental construct. It is totally fictitious, a conceptual identity, but not presence, reality itself, the only true identity. The only one that actually *is*.

Speaking on the basis of the human spectrum, very often what we don't want becomes the underlying motivation to get what we do want, and this involves enormous effort, as our attention is unconsciously completely monopolized by what we don't want. Attention, which is another word for thought, is like water to a garden, so it's actually thought feeding itself. Within the context of this temporal mental construct, it *appears* that there may be some validity to the saying, "resistance creates persistence," as evidence of this can be observed in one's day-to-day experience of living. *Appearing* to function within the context of this temporal dreamlike construct, if one must have a want, it certainly *seems* easier if one simply wants what one *appears* to currently have. The demand and desire for people and things to be different than the way they *appear* to be, *seems* to be the root cause for the vast majority of the *apparent* suffering in this temporal dreamlike mental spectrum.

If there's any possibility of dealing with these dualities it *seems* sensible that it would involve embracing the entire spectrum, but clearly this is not something that can be initiated using one's own personal volition. Any movement in this direction would have to come about organically, and by embracing the entire spectrum one's attention would *appear* to unlock from the aspect of the duality that one really doesn't want to acknowledge or experience. And it is through an embrace of the entire spectrum that a space *appears* to be created for that aspect that one was seeking to manifest in the first place.

Within the framework of this temporal mental construct that *appears* as the human picture, because each aspect of a duality rest on the same spectrum, there is always the possibility of one aspect morphing into the other. However, attempting to have one without the other is an impossible goal, and it *appears* as though it is the goal itself that fixates one's attention on the aspect that one was seeking to avoid. After all, the goal itself is a want, and all wants are driven by the absence of what one *appears* to want. In other words, the goal fixes one's attention on the absence of the goal, and this fixation doesn't allow a space to open for the aspect of the polarity that one would prefer to experience. This is not to say that one can't have ordinary goals within this construct but having an understanding of the nature of how this construct *appears* to function mechanically surely can't hurt. Again, if this doesn't *seem* to arise organically then it would simply be a dialectical process where human thinking feeds on thinking in order to perpetuate more of the same, i.e., more thinking. When one truly sees the reciprocal nature of the two aspects of any duality, then the fear of that which one doesn't want instantly recedes.

Given what we *appear* to be, and that we *appear* to be, within this temporal mental construct, we can't afford to rely on *hope, faith, and belief.* Indeed, those who *seem* to be continuously promoting hope are the ones who *appear* to live with a sense of hopelessness. It *seems* only practical that if one didn't live with a sense of hopelessness, there'd be no involuntary impulse to put a special emphasis on hope. In other words, it *seems* only sensible that one wouldn't emphasize the need for hope unless one felt hopeless. Within the context of this illusory construct, hope *seems* to be a comforter that allows many of us to settle for a blurry vision of a future that never arrives. *Hope, faith and belief* are poor substitutes for certitude. It *appears* as though this is a fundamental human dilemma. Human sense *appears* to always be teetering between choices, flip flopping, due to a lack of certainty. In this context it *seems* one relies on *hope, faith, and belief* when making a choice. It appears that we actually believe that we are separate individuals possessing the ability to make choices.

Speaking on the basis of what *appears* as the human picture, it *seems* as though we have created divisions within divisions, and divisions within those divisions. It *seems* that after all these many years of stressing one side of what are essentially complementary relationships, we are facing a scenario that has reached a maturation point. This *apparency* — is fleeting, passing, never present, finite, measurable. Surely this is not reality, what actually *is*. However, it can — at least in a *seeming* sense — be experienced as a wonderous dream, a play of sorts.

Speaking on the basis of the human scene, all that one — thought — is ever dealing with is thought, it *appears* to be the whole game. It *seems* as though thinking is stuck thinking about thinking, and the problems that are themselves products of thinking. All of the dualistic constructs that *appear* within the temporal human spectrum are designed to ensure the continuity of the pretense that is the one never present temporal-mind itself, and therefore, the characters *appearing* in this spectrum fulfill their roles accordingly. Perhaps by shining a light on this subject, *seemingly* out of nowhere, an approach may spring forth that results in a more sustainable future for humanity. These dualities are part and parcel to this temporal mental construct that is experienced as the human picture. This being the case, it would be nothing short of a miracle if any so-called original approach sprung forth and succeeded in altering the already embedded vibrational structure that has *seemingly* hijacked the entire dreamlike mental construct called humanity.

Given that what *appears* as the human picture *seems* to be dualistic in nature, it makes sense that *the subtle shift,* that was essentially a shift in vibration, triggered another kind of shift, *a duality shift.* The divisive mechanism resulting from *the subtle shift appears* to employ these dualistic mental constructs to ensure the continuity of its pretense, namely, the entire temporal mental construct *appearing* as the human picture. Basically, these dualities are the perfect mechanistic tool, as the word *duality* means, "the quality or condition of being dual." And, the word *dual* means, "consisting of two parts, elements, or aspects." The divisive thinking mechanism resulting from *the subtle shift* separates, divides, protects, and defends. A duality consists of two aspects resting on the same spectrum that are inherently complementary. But it *seems* the divisive thinking mechanism makes it *appear* as though these dualities encountered in our daily round consist of two polar opposites that are not at all complementary, but instead, independent and unrelated. This means the way dualities are seen and understood by human sense is actually due to the interference on the part of an interloper, the divisive thinking mechanism discussed in this book *appears* to be that interloper.

The temporal dreamlike construct discussed in these writings is analogous to a mental movie. This means the entirety of the human picture and its universe would be entirely mental, just energy or thought. Some have labeled it the human mind, the dream-mind, the sensing or thinking mind. It doesn't matter what one labels that which has no foundation in reality. Thought is limited. Life is limitless. Thought is time. Life is timeless. Thought is a movement. Life is alive stillness. Thought is never present. Life is the present, literally. Thought is not being, not present, not now, and thus not, period! Life is *all that is*. Thought is not the opposite of life. Life is now, and now, has no opposite. Life is simply another word for existence. Non-existence is not the opposite of existence, as it's non-existent, it's not here, now, not being, not present to be the opposite of anything. There is no point where existence ends and non-existence begins, because non-existence is non-existent. If non-existence actually co-existed with existence there'd be no need for the word non-existence.

Speaking on the basis of never present *appearances,* it *seems* as though humankind has been grappling with the same dualities for centuries. These dualities *seem* to touch on every facet of human experience, and it can be reasoned that humanity's one-sided approach to these dualities is responsible for the majority of the challenges the human species *appears* to be facing today. The so-called individual, the separate personal self, functions as a duality, in a duality, that *appears* to include numerous dualistic constructs. The twelve dualities identified in this book *seem* to be as relevant today as they were a thousand years ago. Thought *appearing* as separate secondary selves, separates, divides, defends, protects, and assigns labels and definitions to these dualities that touch on every facet of what typically *appears* as day-to-day human experience.

The thought may *seem* to arise, "this all sounds a bit crazy," and that's because it really is whacky, to say the least. It gets even crazier when we consider what we are, or what we *appear* to be, within the context of this temporal dreamlike mental construct, as it is a never present thinking mechanism that is at the root of what *appears* as an insane approach to every duality that *appears* within the human realm. This doesn't mean that the purported temporal-mind is intrinsically bad or evil, because this thinking, sensing, mental mechanism is not an alive, conscious, intelligent entity. This absent interloper has no status in reality. The inequitable approach to handling to the various dualistic relationships within this temporal dreamlike spectrum *appears* to dutifully serve its primary goal of ensuring the continuity of its pretense. This may sound a bit strange, but this relative, conditional, dimensional, temporal mental construct *seems* to have its own rhythm, its own ebb and flow, and given that all of our ideas, concepts, models, and methods have their roots in thought, it *appears* that through the use of thought we have thrown that rhythm totally out of balance. The divisive thinking mechanism that resulted from *the subtle shift appears* to be programmed to separate, divide, defend, and protect, so there really is nothing one can do "personally" that wouldn't add fuel to the mechanism itself. Thought is the fire and the fuel. It's not real, but to itself it *seems* that it's a fire that not only burns ceaselessly but grows without interruption. In fact, it *seems* that it is programmed to keep the fire burning.

The author isn't presenting a method in these writings aimed at instructing a never present personal or lesser self on how to best handle the various never present dualities one *appears* to encounter in one's day-to-day experience of living. There really is no panacea for the *seeming* challenges facing so-called humankind. Any purported approach starting from a limited human perspective would always involve duality. It doesn't matter if one is discussing people in a large metropolitan city or a group of tribesmen in a remote jungle, it's always the same exact mechanism, and the same one *seeming* temporal mental dreamlike construct.

Within this temporal dreamlike movie, that is itself a duality, it *appears* that humans experience a physical world where everything they seek is motivated by, and dependent on, what they personally *think and believe* is its opposite. It *seems* as though one aspect of every dualistic relationship has become more prominent and increasingly problematic. In terms of what *appears* as the human scene, it *seems* there are issues that humankind has grappled with for generations that never *seem* to get resolved. Given what humans *think and believe* they are within the context of this temporal mental spectrum, it *appears* as though one has no actual freedom of choice, and no freedom of action to do anything independent of thought, much less confront these longstanding issues. This never present temporal realm *appears* as a mental construct where thought is simultaneously the projector, the characters in the movie, and the audience.

Actually, and factually, in reality there are no dualities at all. *Only now is.* The now that is now has no polar opposite, and it's not an aspect of a complementary relationship. As already stated, and emphasized, each duality that *seems* to *appear* within the context of this temporal dreamlike construct is made up of two aspects, like a typical coin, and as soon as one wants one side of the coin, one has automatically invited the other. If there was no wanting there'd be no coin at all. Wanting and thinking are synonymous terms, as it *seems* that as soon as one wants one not only triggers the thinking mechanism into another gear, but also, wanting acknowledges that one senses an absence, and this is what *seems* to precede the want in the first place. If there were no recognition of an absence there'd be no want. In this scenario, one's attention is exclusively focused on absence, one *seems* to have a sense of lack, and instantly something is needed or wanted. And who is this one that wants? It is thought itself, as thought is not being, not present, not now, it seeks to be through a process of becoming, a process of "getting there." All wanting involves a future that never arrives because the future can never be present. The want is for something in the future and in the future it remains. It's akin to a total denial of now, reality!

Speaking on the basis of the human picture, is it any wonder that after thousands of years, and thousands upon thousands of so-called innovations, we *seem* to wrestle with the same fundamental issues? For example, it *seems* that we can invent and progress in a multitude of different endeavors, but when it comes to the basic aspects of living, all of our efforts have produced little in the way of results. How long has humankind struggled with unhappiness and depression? How many years has humanity yearned for world peace? How long has humankind *seemingly* grappled with issues related to war, anger, political unrest, divisiveness, poverty, cruelty, racial biases, sickness, greed, selfishness, hopelessness, insecurity, etc.? In spite of our so-called amazing discoveries, inventions, and innovations, we can't *seem* to shake certain issues that nag at the well-being of humankind as a whole.

Speaking strictly on the basis of human or personal sense, *wants, demands, and desires* reveal that one's attention — thought — is on absence, and this is because personal sense *assumes* the two aspects of any particular duality to be polar opposites. For example, only an unhappy person would seek happiness, hence, if you are seeking happiness you are focused on its absence. The word absence is synonymous with lack, nonexistence, unreality, and deficiency. It's not possible for this "you" that you know and experience as your personal self to separate from thought. Basically, the "you" mentioned here would be thought, and being that thought is a movement that never stops to actually be present, never stops to be, means thought itself is synonymous with absence. If one were to speak of the human realm, this may be why it *appears* that many characters in this mental movie have countless *wants, demands, and desires*. It is as though absence — non-existence — is trying to be present through a process of becoming, and it never can quite get there. *Wants, demands, and desires* all involve becoming, trying to fulfill that which *appears* to be absent, but the fulfillment is always going to come tomorrow, and never now, the present, because thought can never be present, it can't be period. Essentially, thought is seeking and searching in futility, running in circles, racing to nowhere. This is the so-called human race.

If one were to speak of the human spectrum, it could be said that the majority of issues that humans *appear* to struggle with are fueled by *wants, demands, and desires,* and it *appears* as though these *wants, demands, and desires* have an activator that initiates their sudden *appearance.* What is it that *seems* to be the catalyst for the sudden arising of these *wants, demands, and desires?* Clearly, if it *appears* that one suddenly has *wants, demands, or desires,* then one must feel that one is lacking, or that there is an absence. Even within this temporal mental construct there's a difference between a want and a need. If most of the characters within the human spectrum aligned their wants with their needs it's conceivable that living would be much easier, but the divisive mechanism doesn't function that way. It's always about more, more, bigger, bigger, better, better, etc. Given that this temporal mental movie *appears* to be continuously moving, shifting, coming and going, ebbing and flowing, busy becoming, but never being, never present, never now, it can be said that lack or absence is synonymous with this mental movie itself. This means as an aspect of the dream itself, any particular character *appearing* within it functions as lack or absence as well, and consequently, if one were to speak of what *appears* as the human realm one can see that the vast majority of people have *wants, demands, and desires*. It makes sense given what humans *appear* to be within this never present mental movie. *Seemingly*, starting under a false assumption, and identifying as a separate little self, a lesser or secondary self, functioning within the context of an inherently fleeting, changing, limited, temporal mental construct, it would be impossible for one to function with a sense of wholeness, so *wants, demands, and desires* would — at least for most — *seem* to be part and parcel to one's so-called human personality.

Humankind's *wants, demands, and desires seem* to correlate directly with their inequitable approach to the dualistic relationships that *appear* within the context of the human picture. Within this time construct, irrespective of the duality in question, most humans *appear* to only *want, demand, or desire* to experience only one aspect of a dualistic construct that is intrinsically two, and this is simply not realistic. If one were to speak on the basis of the human spectrum, everything *appears* to be relative and conditional, so the dualities encountered in this purported construct can *appear* to be only relatively harmonious. But the divisive thinking mechanism that resulted from *the subtle shift* morphs a relatively harmonious relationship into a complex network of issues that *seem* to demand resolution. Speaking on the basis of the human picture, it *appears* as though humankind feeds that which they don't want by over emphasizing what they do want, while thinking about what they don't have, or the absence of what they want. All *wants, demands, and desires appear* to involve a future, in that it's not possible for the future to be present, to be now, all *wants, demands, and desires appear* to be mechanistic mental processes that help to ensure the continuity of thought, as the future never actually arrives, the *want, demand, or desire* is either magnified or replaced. Upon closer inquiry, a want *appears* to be a denial of now, and a clinging to a fictional future, as *only now is.* Now being *all there is,* who or what could deny now? Given that there is no time in now, and that it is never not now, when would this occur? Now or reality, being whole, complete, total, *all that is,* where would any so-called want have status? If life is now, then only now is alive. Life is not alive in a past or future. It's alive now! Starting with the only valid premise there *is* — *only now or reality is* — these questions never *appear.*

The divisive thinking mechanism resulting from *the subtle shift appears* to be programmed to separate, divide, defend, and protect. What is ordinarily called human thinking *seems* to have only one job, and that is to perpetuate thinking in order to ensure the continuity of its pretense. Its pretense *appears* as a would-be temporal mental dreamlike construct that *seems* to be inhabited by separate wannabe selves with separate minds. This purported thinking mechanism is inseparable from its own pretense, so if its pretense were to dissolve or be severed, then its *seeming* status to itself would be annihilated as though having never existed because in reality it has no existence at all, not even in a *seeming* sense. In truth, the present or now is never absent from itself, never vacates itself, there is no possibility of a never present temporal-mind ever having a beginning. Nothing can be done about the so-called human condition, and in truth, there is nothing to be done or undone, and nothing to overcome. Essentially, the divisive thinking mechanism that resulted from *the subtle shift* can't think of a way to free itself from itself, and its many attempts to do so only *seem* to strengthen its own illusory permanence. The thinking mechanism is not a living entity, not aware, not intelligence itself. It can't free itself because it is unaware of its illusory plight. It doesn't want to be free, doesn't not want to be free, it's more like a computer program with a critical malfunction. As the divisions multiply and *appear* to become more refined, this mechanism *seems* to gather more resolve and momentum in its efforts to build imaginary bridges across illusory chasms. This *apparent* movement to connect or unite *appears* to create an unnecessary friction that only results in the *appearance* of more division, and this acts as a catalyst for more thinking and figuring, as it *seems* something must be done to heal these divisions. This process *appears* to go on ceaselessly, it can't stop because thinking can't stop. Even if it could, it would be analogous to a suicide. Thinking is time, and it's not possible for time to experience now, the absolute absence of time. This ensures that it will be forever lack, forever incomplete, forever not whole, but this also guarantees — at least in a *seeming* sense — that there will *appear* to be an insistent and unceasing demand for abundance, completeness, wholeness.

CHAPTER VII

LANGUAGE AND COMMUNICATION

If one were to speak on the basis of what *appears* as the human picture, what is handed down from generation to generation — in the form of knowledge — is integrated into and interpreted through a complex thought structure using whatever language is common to one's geographical location. Regardless of the language, words are merely dualistic conceptual containers. Words by their very nature are divisive and are part and parcel to this movement of energy, this temporal dreamlike construct that is itself a vibration. The temporal-mind *appears* to be playing its own game, and because it's a vibration it functions only in pairs of opposites. These dualities are not really distinct opposites, but words or language makes it *seem* that this is so. Words are inseparable from thought, and in most cases a word points to the thought that *seems* to have given rise to that particular word. The thinking mechanism *appears* to be clever in its use of words and concepts, but only to its illusory self.

When it comes to personal or human sense, one's agreements about most ordinary things is widely known and accepted. For example: a chair is a chair; a car is a car; the sky is the sky; it doesn't matter what language one is using in any particular geographical location. In addition, to those things that are typically believed to be physical forms, we have invented labels for a variety of conditions and emotions, and we have given names to things that *seem* to *appear* naturally on this planet from plants and trees, to the numerous varieties of animals. There *appears* to be no limit to our ability to create and use word concepts to separate and divide, and in a practical sense, within this temporal dreamlike spectrum it *appears* that it can be useful. For humankind it *seems* to have a functional value, it assists us in our day-to-day living in that it helps us to facilitate communication. It can be said that within a limited conditional context, this temporal dreamlike mental construct *appears* to work. Of course, it's always personal, so it depends not only on how one *seems* to be living, one's *seeming* status, but where one is located geographically. When speaking on the basis of what *appears* as the human scene, communication *appears* to be a mysterious process, as one can only interpret and translate through, and from one's personal perspective. It's just not possible to get behind another person's eyes and really know if there's been an exact duplication. This *appears* to be inherent to how the never present temporal-mind *seems* to function. Starting on the basis of human sense, the idea that one can even attempt to get behind another's eyes is built on the assumption that there is an entity there that is entirely separate from another entity. This *appears* to be a clever trick devised by this never present mind itself, because even though each one *appears* to be different to each other, both have been formed using the same ingredients as all the other *apparent* separate entities. In addition, because this never present mind is not aware — as only awareness is aware — it *appears* to ignore the fact that in one's daily round one *seems* to be aware of other bodies, yet one is never actually aware of another awareness. In truth, there is only one entity — if it can be called that — and that would be infinite awareness, reality, the alone one itself, the one right here, now. The divisive thinking mechanism resulting from *the subtle shift* can't be exposed as a fictional, ephemeral interloper, as it is inseparable from the temporal-mind — dream — that thinks up or dreams as what *appears* as the human picture and its universe. Life, reality, now, is all, and nothing genuine will ever be revealed that isn't already present fact to life itself.

Assessing things on the basis of *appearances,* it *seems* that the human brain picks up different things from person to person. There are numerous elements that make up this so-called person you personally know and experience as yourself, but make no mistake, the mechanism operates in exactly the same way despite the purported individual or geographic location. It *appears* as though we are not only stuck in a conceptual world, but we are inseparable from it. Personal sense experiences a world of separation, and consequently, we pretend to use communication in order to build bridges. While this three-dimensional human picture is not reality, it still *seems* that one's daily experience of living takes place within the context of a never present temporal dreamlike paradigm. So, it only *seems* sensible and practical to have a handle on how it *appears* to function within the context of this conditional realm where it *seems* that we live out our personal stories.

Through observation, examination and inquiry, it is possible to deconstruct how basic communication works within the context of this temporal mental movie. There *seems* to be a relatively wide spectrum when it comes to communication. Speaking on the basis of human or personal sense, it can be observed and concluded that when communication between people declines it tends to take on more of a dense or solid appearance, until eventually bullets and bombs are flying, and on the flip side, when communication reaches some point of clarity then it's like one has a good idea of what the other is going to say before they even say it. Theoretically, the cleanest communication within the context of this relative temporal dreamlike construct takes place in absolute silence, and in this scenario, any wanting or seeking to communicate *appears* to be absent. Of course, if one is speaking of the human spectrum, it can be said that for the sake of societies, language has its place, and it *appears* to have genuine practical value. It *seems* that other species on this planet have unique methods of communication as well, but anything said about that would be pure speculation.

Language itself is inherently divisive, and it *appears* to be a necessary division if we're going to have the *appearance* of societies and a diversity of cultures living on a planet called earth. That being said, humankind *appears* as the same one human species, but human thinking takes that which *appears* within this dreamlike construct and seeks to divide it further and further. The obvious divisions are: Mother, Father, sister, brother, Aunt, Uncle, etc., but it doesn't stop there. Human thinking has invented many ways to separate and divide using a myriad of criteria: race, gender, ethnicity, religion, age, height, weight, status, appearance, ability, athletics, politics, personality, human intelligence, etc. It *seems* it would be obvious to even the most unbiased observer that this process of dividing and drawing lines has reached a maturation point in some societies. Starting from a divided perspective, it's not possible to avoid creating further divisions. Humankind's ability to draw lines and create division *appears* to have no limits. It *seems* that since *the subtle shift,* the demand to divide and separate is relentless, and it *appears* that this has had wide-ranging and profound consequences. If one were to speak on the basis what *appears* as the human picture — the way things *seem* as reported by the human senses — it *seems* we have reached a point where our humanity is buried beneath a bloated complex structure of words, labels, ideas, concepts, interpretations, and invented definitions. These illusory divisions are incorporated into this relative dreamlike mental spectrum, and they are used to generate even more divisions within an already bloated construct of preexisting divisions. It's just one small aspect of a robotic like divisive mechanism that *apparently* has no off button.

Speaking on the basis of human sense, language is the basic method of human communication usually consisting of spoken or written words that can be voice sounds, symbols, or gestures. Language is employed to express thoughts, feelings, and emotions. Often thinking is so automatic and mechanical that like automatons we are verbalizing without knowing what or why we're saying what we're saying. Even the numerous ideas, thoughts, and feelings we have about ourselves, about what and who we are, have their roots in thought. It's not possible to interpret a feeling without the use of thought in the form of knowledge. In essence, the self you are familiar with is inseparable from thought, so there is no individual self that is sitting around thinking up thoughts. The personal self you're familiar with as yourself is the temporal-mind's self, and it maintains this illusory identity through the use of thought. One can notice that there *appears* to be a dialogue that's going on without a moment of stoppage. Most of this dialogue is concerned with thoughts about this little or secondary self that you *think and believe* is you. Thought is always trying to ensure its continuity and pretending to be your personal human mind would just be another activity of a never present dreamlike mental construct. It *appears* as though there are thoughts, but no one is present actually having them, it's a mechanical process. Thought will surely seem to argue this point. I'm not going to speculate, but it *appears* that there's no one at the helm, and it *seems* almost impossible to say anything about thought without the knowledge — thoughts — we *seem* to have about thought. This *appears* to be on display more blatantly in human relationships, particularly when one person is disagreeing with another person defending his or her specific viewpoint. Often people *seem* to be totally blind to the fact that although they disagree with the person that *seems* to be in front of them, they are surely in agreement with someone else, and it is not actually *their* unique point of view that they are energetically defending. The illusory viewpoint they're protecting and defending, is not really *their* viewpoint, as they have no way of knowing the source of the information, nor can they even be certain if it is correct. Moreover, upon closer examination, their entire argument is built on the assumption that it is possible to really know, in the purest sense of the word. On the surface it's amazing that people protect and defend their viewpoint as though it is exclusive to them, and they ignore the fact that it's not so unique and that they simply agree with someone else's pre-existing viewpoint. They present their point of view under the guise that it is theirs, when in fact it was hijacked again and again. All personal viewpoints are a collection, or a grab bag of other viewpoints, because the information that acts as a resource for any individual viewpoint did not actually originate in the person proclaiming, "this is my viewpoint." That is to say, the grab bag may *appear* different, but the items in the grab bag are not all that original. If the contents of the grab bag are not original, then the grab bag itself is not original either. This picture is more convoluted than it *appears* when you consider that both parties essentially face the same dilemma. This illustrates how the divisive thinking mechanism resulting from *the subtle shift seems* to function. It's continuously generating divisions, adding more and more layers, and reinforcing the illusion of separation. This may sound somewhat crazy, but keep in mind that each *apparent* separate self is inseparable from the mechanism and has absolutely no freedom of action in this scenario. This is yet another *seemingly* clever manner in which this never present temporal-mind ensures its continuity.

Speaking on the basis of the human picture, it *seems* as though we've reached a point where every word that we use came from someone else, every thought came from someone else, and even every experience has been experienced before, and then interpreted through the use of the same words that we ourselves use in our daily experience of living. Feelings are thoughts as well: feeling good, feeling bad, feeling happy, feeling sad, feeling hot, feeling cold, etc. If you have a feeling of sadness then you have all these references in the form of thoughts that tell you so, and these thoughts are basically just a database of information and experiences. Is there a single person *appearing* on this planet today that can define and describe themselves without relying on the knowledge and information that has already existed prior to them arriving on the scene? Of course not. What this really means is that within this temporal dreamlike mental construct, it's really not possible to function as an individual that is entirely separate and apart from others. Not only are words dualistic conceptual containers, but they are inextricably tied to thoughts, and thought is inherently divisive, so language separates and divides. If one were to speak of the human picture it's impossible to deny the fact that the human database of words and definitions has grown into an incredibly bloated structure. For example, about 1,000 years ago, what is called Old English had approximately fifty to sixty thousand words. There never *seems* to be a total consensus on this sort of thing, but as I write this book today, it's estimated that Modern English has close to one million words. Given that human thinking is a divisive mechanism, and since *the subtle shift*, it *appears* as a divisive mechanism on steroids, this *seems* to be reflected in the bloated and complex structure we call the English language. And this is true of other languages as well. Language itself *appears* to be useful, if not necessary when it comes to facilitating communication. In one's daily experience of living, one would be hard pressed to function in its absence. Certainly, we could do quite well with far fewer words, but the divisive mechanism that resulted from *the subtle shift* is relentless when it comes to separating and dividing. Language *appears* necessary within the context of this mental construct, but through the use of word concepts we constantly separate ourselves, not only from each other, but from the world that *appears* to be around us. These divisions are obvious, as they are inescapable in terms of our day-to-day experience of living. If words were simply used to assist us in basic communication, perhaps things would be far different. Given what we *appear* to be within this mental movie, as soon as we say "I," "me," or "you," it *appears* there is division. As soon as we identify ourselves on the basis of race, ethnicity, politics, or religion, we have planted a divisive seed that must bear its fruit in the form of conflict.

Speaking on the basis of what *appears* as the human scene, language can be a useful tool in our day-to-day experience of living, but it *appears* there is no way to harness the mechanism that *seems* to employ this tool to ensure the continuity of its pretense.

CHAPTER VIII

LIFE VS LIVING

What do we personally really know about life? We *appear* to know an awful lot about the content, and that *seems* to be reasonable considering the fact that we invented the labels and definitions that we use to interpret the content, but the so-called content is not life itself. We *appear* to know a lot about things, but do we really know anything? As humans it *seems* our knowing is within an inherently conditional and limited framework. When it comes to the content — what we claim to know — as one would expect, there *seem* to be as many finite opinions as there are finite personalities. Like characters in a movie we *appear* to know things within a temporal mental movie, and hence, we *seem* to be relegated to the conceptual, not the actual, not reality itself. Humanity and dream are synonymous terms, and accordingly, dream, dreamer, purported characters, countless finite forms, and even thoughts and emotions are part and parcel to this never present dreamlike mental construct. This entire thought package that *appears* as the human picture has a commonality, all of it *appears* within the context of never present time. The temporal-mind experiences or senses a world that is its own projection. It fully believes in the reality of this thought projection. As a character in this temporal mental movie you are part and parcel to this mind. Indeed, this mind is posing as your personal self. It's not *your* true identity but given how this mind *seems* to function it's unlikely that you would believe otherwise.

As characters *appearing* within the framework of this temporal dreamlike construct, we *seem* to live with the false assumption that there are individuals separate and apart from each other, and the world at large. We *seem* to function in an agreed upon spectrum that has been projected and solidified by a flow of information that *seems* to be endless. We pass down what we call knowledge along with our beliefs and experiences from generation to generation. This process of growing and expanding this global human database of knowledge, beliefs, and experiences, and then handing it down from generation to generation relentlessly continues on and on, like a machine stamping out parts. It must be said that in a conditional context it *seems* to have a degree of workability. At a practical level the machine works, at least to its own benefit. Speaking strictly on the basis of the human picture, it *seems* to work in terms of supporting basic communication and helping one to function relatively intelligently in one's so-called society. From a superficial perspective there *seems* to be real value in possessing this knowledge. From the perspective of human or personal sense it *appears* to be invaluable, and it enhances our ability to be what has been predefined as successful in this world, and it *seems* to be very necessary as a facilitator in various communication circles. In other words, a certain amount of knowledge is essential in order to make our way in this world, but when it comes to questions about life, it doesn't *appear* to have any value at all. Speaking on the basis of *appearances,* it *seems* as though we're trapped in a world of ideas and mental constructs that we cannot escape because we are inseparable from them. In fact, the more individual effort we exert to break free, the more we strengthen this illusory prison. But then exactly who is it that is trapped in this prison? You as you know and experience yourself are not separate from thought, and it is this "you" that *appears* to be trapped in a prison, but this is a clever trick on the part of the mechanism itself. If there is anything there that can be called a prison, this is it, as both prison and prisoner are one in the same. It's an illusion the temporal-mind uses to perpetuate the continuity of an illusory world it purports to experience. If this "you" tries in any way to figure a way out, a method of escape, then the illusion that *appears* as the prison itself is further reinforced and solidified.

Within this world where it *appears* there is a "we" that proclaims to experience a fleeting, changing spectrum, it's difficult to make any definitive statements about anything at all without putting some kind of conditions in place first, because this *apparent* reality is relative. It's rooted in a field of time, and all conditioning requires time. In the absence of what has been *seemingly* handed down to us by others, and in the complete absence of the individual indoctrination, conditioning, and what our brains have picked up along the way, what do we really know? In a peculiar way we use vibration (thought) in the form of what we call memory, knowledge, and information in futile attempts to understand ourselves — what we are, who we are, where we're going, where we came from. It's analogous to a game where a fictional self is working in futility to know and understand a fictional self, because this so-called we, you, me, I, them, etc., are part and parcel to the same vibrational mental construct that *appears* as the human picture.

If one were to speak of one's *seeming* daily experience, people often say, "I'm working on my 'self'." One quite reasonably might ask, who is this self you are working on and who is the one doing the work? Starting off from a divided perspective only results in the solidification of that division. Is it divided because it's a vibration, or is a vibration because it's divided? One *seems* to be left with two co-dependent aspects that really have no basis in reality but *appear* to give presence to each other. It's analogous to a coin, each side gives credibility to the other. Only if one had never picked up the coin would one be free, and this is actually the case. In reality, the alone one, life itself, never picked up a coin with two sides. Having no opposite, life is eternal completeness, wholeness, totality being *all there is.*

Surely what *appears* to have resulted from *the subtle shift* is insanely stealth-like. Why wouldn't it be? It's actually inseparable from a never present temporal mental continuum. All of our thoughts, feelings, and experiences *seem* to be interpreted through the filter of a complex thought construct, it's like an enormous database of knowledge, thoughts, feelings, and experiences that continues to expand and gain momentum as it builds and feeds upon itself. So, our thoughts and feelings, and what we *seem* to experience, is really part and parcel to the activity of the never present temporal-mind. Basically, it *thinks and believes* that it is you. It doesn't know that it is a pretense. It doesn't know that it and the world it experiences is not real, not reality, not present. The never present temporal-mind is both belief and believer, so it's no surprise that it experiences a realm of belief. As a consequence, the characters *appearing* within the framework of this mental construct create experiences and interpret everything through a sort of belief package. The temporal-mind is both belief and believer, dream and dreamer, lie and liar. It has no basis in reality. In your *seeming* daily experience it *appears* that everything is what "you" believe, and if you don't believe that, then that's what you believe.

Given what we *appear* to be, our dilemma, if there is one, is not associated with life, but with our day-to-day experience of living within what *appears* as a three-dimensional temporal spectrum. It *seems* as though we're always wrestling with how best to live with each other and the numerous sentient creatures that *appear* on this planet, while at the same time searching for ways to address the problems, conflicts, and limitations that *appear* to be inherent to this illusory never present dreamlike spectrum. It is thinking that *appears* to separate us from each other and the world that *appears* to be around us, because it is the temporal-mind that is thinking, dreaming, or projecting as what *appears* as the entire human picture.

What *appears* as the human spectrum and its universe is all but synonymous with lack and limitation. This is not stated to be negative or to ridicule what *seems* to be one's day-to-day experience of living, nor does it mean that it's bad or evil. If one were to closely examine what *appears* as the human picture, one finds lack in virtually every aspect of so-called human experience, and limitation *seems* to be inseparable from this never present mental construct. The question may arise. Why is this the case? The temporal or human mind's world and its experience of it invariably involves observable or measurable forms. Whether *seemingly* physical or mental, forms have limits, and they *appear* to have a starting point and an end point, so it can be said that they are measurable. It may be a sight, a sound, a smell, a taste, something touched, or it may even be a thought or an emotion. Everything this mind experiences is noticeable or observable, and hence, measurable. Everything it experiences is not only fleeting, transient, but limited or finite as well. All of it *seems* to *appear* within the context of passing time, and given that time is never present, these *seeming* forms are never present, so it can be said that they lack presence! Having no presence would be the embodiment of lack. If something is not even present, how can it be real or true? Only the present is present, hence, if something is not present, not here, now, it is non-existent.

What *appears* as the human spectrum and its universe can just as easily be called the realm of not, as it is not being, not present, not now, and therefore, not, period! Personal or mortal sense starts from lack, with lack, in lack, indeed as lack, and functions as lack, in a world of lack, a realm of not. It clings to *hope, faith and belief* for fulfillment in a future that never arrives. It's always going to become or be something tomorrow, get something or somewhere in the future, never can it be fully satisfied now. For personal sense, the future holds its dream for fulfillment and satisfaction. It *seems* to live in hope and dies in hope. It's really not possible to discuss what *appears* as humanity — dream, the realm of not — without discussing the subject of lack. The word *lack* means, "the state of being without, or not having enough of something." The word lack is synonymous with absence, want, need, deficiency or scarcity. The word *limitation* means, "a restriction, a defect or failing." Surely the most common problem facing humankind *appears* to be lack and limitation. In every aspect of human experience there *appears* to be a need, a want, a shortage, a call for more, a demand for bigger or better. Speaking on the basis of personal or human sense, one might say the entire economic system is based on the concept of lack. Companies are busy scurrying about trying to fulfill a want, a need, or a desire. Often, they're the ones informing us as to what it is that we need, or perhaps better said, they inform us that we are lacking the particular product or service that they're currently promoting. If this issue of lack could be resolved by seeking answers in the so-called human paradigm, the realm of not, then it *seems* only sensible that it would have been resolved long ago. Given that it *appears* as a realm of not, all answers are "not" the answer.

Consider any aspect of purported human experience and the issue of lack *appears* to have status. In human circles there is talk of the haves and the have-nots, but even the haves never *seem* to have enough. There is a widespread mantra inseparable from humankind that purports need and want at every turn. The basic message is one of lack. The expanding list of humanities' wants, needs, and demands *appears* to be the only thing that is not lacking. One can't get rid of lack, as it is not present to be gotten rid of. It's amazing how hard humans have worked trying to resolve this issue called lack.

Lack and limitation are inherent to humankind. One could circle the globe and meet people from all walks of life and develop a list of lacks and limitations that would *seemingly* go on ad infinitum. A lack of: money, friends, lovers, health, peace of mind, freedom, time, patience, sleep, security, resources, opportunity, education, communication, power, understanding, interest, agreement, control, vision, strength, popularity, rights, optimism, conviction, morality, empathy, commonsense, ethics, tolerance, stamina, solidarity, restraint, harmony, desire, action, honesty, sincerity, depth, energy, attention, comprehension, flexibility, enthusiasm, stability, closure, certainty, completion, confidence, confront, rhythm, flair, support, compromise, openness, creativity, originality, transparency, and the list goes on ad infinitum.

The list of lacks and limitations that *seem* to haunt man from the cradle to the grave is exhaustive indeed, but there is one lack that *seems* to be the basis for all of the others. Humankind lacks presence, and no amount of thinking, effort, focus, exertion, hope, faith, or belief can resolve this dilemma that *appears* to be inherent to humankind. Always moving with the flow of time, humankind is never present, never being, never now, never *what is,* and hence nonexistent. It is the absolute embodiment of lack.

Irrespective of how things *appear* and speaking on the basis of what appears as the human picture, every purported person's destination is exactly the same. Of course, the beliefs and myths regarding what happens after what has been defined as clinical death takes place are numerous and diverse. From a human perspective it doesn't matter what theoretical story one believes, as one is left functioning as lack, in lack, and ultimately one *appears* to become the epitome of lack. One is gone, dead, lacking even the *appearance* of being a character in a never present dream.

What *appears* as the human picture is vibration, and therefore, it *seems* to only function in pairs of opposites. It's no wonder that having a harmonious relationship is rather challenging for so many of the *seeming* characters within this temporal dreamlike construct. The never present temporal-mind's thinking or dreaming would be the always changing, shifting, moving, coming and going of what *appears* as the entirety of the so-called human picture, i.e., air, water, fire, bodies, forests, disease, war, houses, cars, planets, etc. This never present temporal mental construct lacks stability, as it's a continuously reacting vibration, *appearing* as numerous forms that are always changing, passing on, leaving, never really arriving, never actually being. In one's daily round this would be how things *appear,* but if there was no life, no presence — the formless, or whatever word you care to use to mean, that which simply *is,* that which never passes or changes — would it even be possible to notice that the forms *appearing* within the context of human experience are fleeting, basically, ephemeral in nature? It *seems* unlikely as nothing would be, there would be nothing being, nothing here, now, no reality, no presence.

In the *seeming* human construct, change *appears* to be the only constant. Surely this isn't reality or life, what really *is*. Ask yourself as to the nature of reality or life. Would that which word reality or life stands for be transitory, changing, shifting, leaving, and never really arriving? Can it be said that reality or life is never present? Is reality a dream, a lie, or a hallucination? Is reality, now, the present, subject to the effects of time that is never present? Does *that which is* — timeless now — age or decay?

Life is reality! Life is now! Only now is alive! Only now is, so it can be said that only life is! Is it ever not now? Never! All is now, and now is what all is. Don't take the author's word for it. Only you can address these statements. If one were to simply believe another, then it *seems* one is relegated to an experience of living in bondage to a conceptual construct. Often the so-called human approach is to believe or follow another. Human or personal sense is relegated to the conceptual and not the actual — that which truly *is*. This means human worship is relegated to worshipping and living in bondage to mental concepts, ideas, and beliefs. This is starting with a false premise, under a false assumption, that results in a fallacious conclusion. So, what we *appear* to have today on a human basis, is numerous doctrines that promise salvation, liberation, revelation, enlightenment, etc., yet they share a commonality in that what is promised is never delivered now, it's always promised to come in a future that never really arrives. All of these so-called doctrines have their roots in human thinking. While thinking *appears* to be a process that involves change (time), being is not a process at all, being is changeless, the total absence of time, being simply *is*. Being is already being, not a state that is yet to come. Being is being *all that is* being. There is simply nothing else, no "other" besides boundless being.

CHAPTER IX

DISHARMONY

If one were to speak on the basis of *appearances,* why does it *seem* there is so much dissonance in our world? The temporal-mind that *seems* to be thinking or dreaming as the human picture is vibration, and vibration is the embodiment of duality, so conflict is inevitable. The entire so-called human spectrum where disharmony *seems* to *appear* is nothing more than dreamlike thought, not ever-present reality. The never present temporal-mind is dreaming as this world, inclusive of all that it *appears* to contain, i.e., bodies, planets, air, water, fire, trees, space, time, houses, cars, etc. Given that this *apparent* dream that is only experienced by the temporal-mind is not actually real, means we don't really need to think smarter, better, sharper, with more clarity, nor do we need to think deeper. More of the same poison is not a remedy, not a cure for our problems, because all thinking would be the temporal-mind's activity. There is no personal you and thus no personal mind to do any thinking. This means in our day-to-day experience of living thinking can't help us to resolve difficulties. If there is no personal self, there can't be any difficulties. Starting under an assumption rooted in personal sense, it *seems* one is left thinking about ways to resolve issues that are themselves products of thought, and hence, inseparable from thinking. Temporal or human thinking *appears* to separate, divide, and experience a world where dualities are the norm. Hot and cold, up and down, happy and unhappy, wet and dry, peace and war, etc. This is how the vibration *seems* to function.

It *appears* as though all of the energy and effort exhausted to know and understand and to figure it all out just adds fuel to the fire, creating more conflicts and disharmony. It's entirely plausible that the never present temporal-mind functions in this way in order to ensure the continuity of its pretense. In other words, this is how it creates the illusion of permanence to itself, and then experiences it as such. To life, now — *that which is* — is any of this purported dream going on? Nope! There is no prior to now, no after now, no "otherness" at all. Just one alive conscious presence is *all that is present*. There can be but one all, one mind, that being infinite mind, the alone one, the sole self, consciously present here, now.

Humankind *appears* to be in a futile race to somehow, some way, reconnect, join, or merge with the all-ness that is life itself. There *seems* to be a void, an inherent sense of separateness, and the demand to resolve this issue is the fundamental factor driving humanity, yet no one *seems* to be aware of this, and to be perfectly honest, there's no valid reason why anyone should be. It's not possible to separate your sense of self from thought, as you as you know and experience yourself are totally inseparable from thought and thought *seems* to only be interested in its continuity. It *appears* there is a survival mechanism built into the human body itself, but the so-called human body is part and parcel to the same one temporal mental movie that has no foundation in reality.

If what *appears* as the human spectrum is nothing but mental or thought, which never really *is,* then the end of thought is the end of you — at least you, as you personally know and experience yourself. If one were to assess things based on human or personal sense, thinking, behaving, acting as thought, it is completely in your interest to do that thing which ensures your survival. Oddly, both thinker and thought are one in the same, and I'm not asserting that there's anything wrong with thinking. I don't want to speculate, but animals studied in their natural habitats *appear* to experience some form of thought. But undoubtedly what is ordinarily called human thinking *appears* to have mutated into something quite insidious, and how we employ it has manifested into a threat to life as we personally know, define, and experience it.

Speaking on the basis of human or personal sense, what is it that ensures your personal survival — not as the physical body that you likely *think and believe* you are, but as thought, the so-called mind? And does it *seem* you are enhancing or disrupting the harmonious functioning of the body through the use of thinking? For all we know the vast majority of our ideas concerning the human body may be doing more harm than good. People point to the fact that it is through our thinking that we have been able to increase our ability to survive and to live longer, at least as a body. Perhaps that is the case, but at what expense? And is longer actually better? If it is better, is it better for the entire planet? Has it been better for the countless other sentient life forms we *appear* to share this planet with, or are we throwing everything completely out of balance? It is interesting that we spend billions upon billions on research, looking for better and better ways to treat diseases, while at the same time we spend billions upon billions researching and developing the most effective ways to kill people. There is something wrong with this picture if you ask me, but it *seems* we go merrily along never questioning this *apparent* insanity. There's an obvious cognitive dissonance here, and again, no one *seems* to mind. I hear people say, "Oh well, what can you do?" Perhaps there's nothing we can do. Maybe there is something we can do. Maybe we're already doing too much. I really don't know, but it does *seem* we have reached a point where we should at least look at this issue with a certain ruthless honesty, or we will likely pay a heavy price for our refusal to address what *appears* to be the fundamental issue confronting humankind as a species. Can we come to terms with what we *appear* to be as a result of *the subtle shift?* Indeed, can we come to terms with what we *appeared* to be prior to *the subtle shift?* If the divisive mechanism discussed in this book were to vanish into oblivion, would we — speaking on the basis of what *appears* as the human picture — bloom into something extraordinary? I really don't know. I have a sense that it's distinct for each organism, because if there's anything close to being unique or original in this construct, it is the human organism and not the *apparent* human being. If one really *appears* only as a character in a temporal mental movie, perhaps the movie would *seem* to have a better ending, but it would end, nonetheless. What *appears* to be born in time must end in time.

A nice dream is better than a nightmare, but reality precludes both!

If we can face what we *appear* to be, indeed, that we only *appear* to be, within this limited temporal spectrum, head on with courage, with ruthless honesty, then perhaps nature will initiate a shift or realignment, but then, nature itself is part and parcel to this mental construct. What other choice do we have? We can sit back and think about it, try to figure it out, come up with more solutions to our illusory problems that have manifested as a result of our refusal to confront a fundamental problem, *the subtle shift,* for it *appears* to be where the human species went off the rails. If this experience we *seem* to be having as humans is nothing more than a dream, a mental thought movie, and we are just characters in said dream, it *seems* only sensible that if one is having a beautiful dream there is no point in waking up, as waking up in this context is the end of us as we know and experience ourselves. If on the other hand the dream has become a nightmare, the movie has morphed into a tragedy, then it *seems* that it not only demands, but also deserves inquiry. But does it really? If the human picture *appears* as a nightmare then it would be the temporal-mind thinking or dreaming as a nightmare. The temporal-mind discussed in this book doesn't actually exist, as it is not being, not present, not now, it is not, period! It's really not possible for *that which is not,* to fix, correct, heal, or improve a mental movie, a dream world that is not, in a place that is not, at a time that is not.

Given that we *appear* as characters within a temporal dreamlike mental construct, it *seems* that we have no choice in this scenario. It *appears* that the divisive nature of never present thought will continue feeding an already bloated never present structure. In other words, from a personal standpoint we'll get more of the same, and the divisions will intensify, and the result will be an irreversible conflict within the framework of this temporal dreamlike mental construct.

The demand to control everything has resulted in many of us feeling like we have no control, and that each and every one of us is ultimately helpless. As odd as it sounds, it *seems* that if man was wiped from this earth, the planet would then go through some type of natural cleansing, and perhaps a healing of sorts would take place. As a species we're not really the problem, but human thinking is like an uninvited guest acting as the head of the household. Even the planet *appears* in the context of this temporal dreamlike mental construct, hence, the planet as it *appears* to us is part and parcel to this dreamlike movie itself. This means our sense of the world, our entire experience of the world and how we perceive it, has no foundation in reality. It is nothing more than a dreamlike mental construct that we *seem* to experience as reality. Even the so-called "we" is part of a never present temporal-dream. The temporal-mind purports to experience its own mental movie, or time dream, so in a personal context there's nothing any so-called individual can do, as the only or sole individual would be reality, presence, or whatever word that means to you *all there is of all there is,* and not some transitory, passing, fleeting, never present dream character. In other words, in very simple terms, all is good even if it doesn't *appear* that way, as the so-called human senses are not able to behold that which *is,* reality, that is everywhere present and nowhere absent.

CHAPTER X

PARROTS AND PUPPETS

Speaking on the basis of personal or human sense. Do we have freewill? Is there really someone at the helm? Do we have freedom of action? Can people really change at a fundamental level? Do any of our well-intended systems, methods, -isms, and -ologies aimed at human behavior do anything at all, or given that they are all products of thinking, are they nothing more than empty promises, would-be aspirations, and repetitive admonishments? I will offer my opinion pertaining to these issues, but I freely admit that they have no more or less value than opinions offered by any other characters *appearing* within this temporal mental construct.

Speaking on the basis of what *appears* as the human picture, who and what we *seem* to be today *appears* to be a combination of DNA, conditioning, environment, family history, neurochemistry, education, plus the knowledge and information that has been put into us. There may be unknown elements as well, but these are the relevant ones in regard to our discussion here. As a result of *the subtle shift* it *seems* human thinking has gone off the rails and is as they say, "large and in charge." It *appears* to function like a computer software program, so it wouldn't be wise to associate freewill or freedom of action with a mechanistic structure. What is ordinarily called human thinking is a never present temporal-mind. It's nothing but vibration, energy patterns, and no one is really at the helm calling the shots.

What about the human body? Within the context of this temporal dreamlike construct are you this body? Did not the body *appear* here in this construct before a name, more accurately before your name was put upon it as a label? That body could have been given any name, so it is somewhat arbitrary. Odd as it may *seem,* even the word body is nothing more than an invented label. Taking this a step further, do you say that it is your body? From this perspective, if this is the case, it *appears* there is a "you," and there is a body that belongs to you. Many characters that *seem* to *appear* within this temporal dreamlike spectrum *seem* to *think and believe* this way, while others feel their identity literally is the body. Still, there are yet others that *appear* to cling to various spiritual and religious theories, beliefs, and ideologies. Some believe they have a soul; some believe they are spiritual beings having a human experience. Others believe they are the product of a creator. All of these so-called positions *appear* to share a common thread. All of them rely on a movement of knowledge, ideas, and concepts, that have *seemingly* been handed down from generation to generation within the context of passing, never present time. If you were to ask someone for their personal opinion in this regard, they would express to you what they personally think. What a so-called person thinks, is dependent on various factors: family history, education, conditioning, societal and environmental exposure; even genetics may play an important role. While behavioral genetics is focused on comparing the genes of people with different abilities, there are researchers in the field of evolutionary biology that concentrate on comparing the genes of different species. I'm not going to explore these speculative theories and studies here, but I will say that DNA may play a central role in humankind's ability to communicate using a language that *appears* to be more complex than that of the other species *appearing* on this planet. This doesn't automatically mean that as a species we are more precious than all the other species on this planet. If DNA does play a significant and prominent role in our ability to use complex language — then at least within the context of this temporal dreamlike construct — *the subtle shift* may be much more than a theory, as language and thinking are inseparable, and what we call genetic mutations do *seem* to take place from time to time.

You may be asking how does all this pertain to the title of this section, Parrots and Puppets? If one were to speak in terms of the human picture, given your particular upbringing, environment, genetics, family history, neurochemistry, DNA, conditioning, and the knowledge put into you, is it possible that you could think, feel, and behave any different than the way you *appear* to? The answer is a resounding no! Speaking in this context, it *seems* that we are all to a great extent parrots and puppets. Fundamentally, in a so-called personal context, we as we know and experience ourselves are simply characters in a temporal mental movie. The so-called human mind is a temporal or time-mind, and there only *appears* to be one never present temporal-mind, and not numerous separate minds. This temporal-mind *appears* to involve the so-called human senses, thinking, emoting, imagining, even memory, all of which are fleeting, passing, changing, never present processes. There is only one *seeming* temporal-mind, and this appears to be a blessing, as it can be concluded that there is no rational reason to go out and about in the world attempting to resolve the many problems that *seem* to demand attention. It *appears* one is only dealing with only one never present dream, one concedes and acknowledges this fact, then goes off and does what's practical and sensible, minus the baggage that would ordinarily go along with daily human experience.

The *seeming* temporal-mind thinks or dreams as the human picture and the entirety of its universe. It acts as the projector and the *seeming* experiencer of its own projections. If one were to speak on the basis of *appearances,* it *seems* that genetics play a prominent role, as they are the starting point when it comes to what humanity *appears* to be today, at least as it is defined and interpreted by human sense. Our individual behavior *appears* to be predetermined and programmed, and most of what we speak of is simply the process of parroting what has been put into us; what we've heard from someone else, read at some point somewhere, or perhaps even acquired through genetics. We even go so far as to mix all these things up and then, put them out declaring them to be something new, something completely original. We take fragments and parts from other so-called viewpoints and then say this is my viewpoint. It's just not honest to say this all belongs to one individual that personally came up with this in its entirety. It's an absolute lie, but there is persistence to this, as this is the nature of a lie, and upon closer observation it can be concluded, based on *appearances* only, that there is just the mechanism itself — the lie — ensuring the continuity of its pretense, it *appears* as both the lie and the liar, the belief and the believer, the dream and the dreamer. To now — reality, the alone one — none of this is going on at all, it is not existent, as *that which is, is all there is,* leaving nothing else besides. The entirety of human experience can be likened to a typical human dream. It's not a perfect analogy, but it makes the point. In an ordinary human dream none of the characters *seem* to be aware that it's a dream, all types of crazy things can *appear* within the framework of a so-called human dream, but upon awakening it's like none of it ever occurred. All of the characters that *appeared* to be real in the dream are nowhere to be found, even the one that pretended to be you. Upon waking up from a purported dream the only one left is you, the *apparent* dreamer. If one does some exploring, it can be seen that one didn't actually decide to have a dream, you didn't control all that *appeared* to happen in the dream. The so-called dream never really happened, never really occurred, so you wake up — sometimes in a panic, depending on the type of dream — only to realize that it was all fake, untrue, not real, a mere passing dream, a sort of mental delusion. Nothing actually happened to "you."

CHAPTER XI

RELIGION

Assessing things on the basis of what *appears* as the human spectrum, it *seems* as though religion itself reveals its own inherent impotence when it comes to dealing with the complex issues of humanity. It starts off from a divided perspective and quite naturally ends up generating more division. Any theory, discipline, method, -ism, or -ology that promises freedom or a release from the burdens that *appear* to be part and parcel to functioning as a character within this temporal mental construct have no basis in reality. These so-called approaches all have their roots in the same *apparent* false assumption. They all *seem* to suggest that one must "become" or "get there." This book is not aimed at negating or devaluing the many doctrines, approaches, and methods that *appear* to dot the human landscape. This work merely points out that they are neither good nor bad, but simply not, as they *appear* within the context of time and time is not being, not present, not now, not, period. As it is never not now, what is not now is not — non-existent.

The numerous religious, spiritual, metaphysical, so-called new age, psychological, and philosophical approaches making claims, making promises and guarantees, are completely spurious at best and without merit. Without exception they all start with a false premise, a fictitious assumption, hence, they start as nothing, do nothing, and end in nothing, indeed as nothing. The author isn't attempting to negate the human experience. When speaking as to the nature of reality it's never about the words. Words about reality are not reality itself. The author is merely pointing out that the reader reading these words is not a temporary, fleeting character functioning in a temporal mental movie. The purported self that *seems* to be fooled by *appearances* is not the self that you be, right here, now. The entirety of what *appears* within the context of this temporal mental construct that *appears* as the human picture and its *seeming* universe has no foundation in reality. As time is *what is not anymore or what is not yet,* time is literally not. It is non-existent. Reality knows neither time nor timelessness. Reality is not a world of opposites, as it doesn't co-exist with another. There can be but one reality, one now, one presence, one all, one infinity, and this is not in a numerical sense. This simply means that when it comes to existence, reality – what really *is* — there is nothing present unlike itself throughout its infinite present-ness. If there's any sense that the author is negating the many works of humanity, the author suggest that one remember that it's not possible to negate what is not present in the first place.

If one were to speak of the human picture, it *seems* as though the temporal-mind has an astounding ability to interpret and translate experiences, and then turn them into something more profound and meaningful than they really are. Over time, all of these thoughts and interpretations resulting from these many experiences evolve into complex energy patterns, thought structures that *appear* to be imposed on others, sometimes it *seems* they are used to control and manipulate the masses. Such is the nature of human thinking.

This may sound strange given where we *appear* to be today, but there's just no way we're going to mass produce a Buddha, a Jesus, a Moses, a Muhammad or anyone else for all that matters. Whatever it is that occurred in these specific cases, if anything, no one really knows. It's likely they had no way to comprehend what happened, if anything at all. I'm more intrigued by the fact that they're all men, and no one *seems* to take issue with that fact. It's no accident that they're all men when we look at how dominant men have been in societies over the years within the context of this never present temporal mental construct. In the so-called human picture it *seems* that there have been those who proclaim there was a fall from grace, a separation from source. In other words, this is not reality itself attesting to this fiction, all of these men along with their propaganda have been affirmed and confirmed by others, by personal sense. All of these purported messengers, saviors, or deliverers, if you will, represent a fictitious perspective that suggest one must get there, become more present, unite, enter the now, or reconnect with source. Many even make profound proclamations about their particular god. Some declare that god is all, however, there's always a but, another besides god. They simply can't concede to reality. God is all or there's no god at all, and the term god would be just another word concept, another attempt to contain that which is utterly ineffable. All of this activity *seems* to be the work of a never present interloper, the never present temporal-mind attempting to steal presence, as it has none of its own.

The temporal-mind that thinks or dreams as this temporal mental dreamlike movie, is simply vibration — energy — that is always moving, shifting, changing, reacting, and passing on. This purported temporal-mind is never present, it's an absent interloper, it is not being, not present, not now, and thus, not at all. This poser is just vibration, an energy pattern that thinks or dreams, projects and experiences a temporal mental movie. As vibration, an energy pattern, this imposter *appears* to have continuity, but only to its never present self.

Assessing things on the basis of human sense, regardless of the purported saviors of humankind, we *appear* to be left with the *seeming* mess we have today. There's no point in condemning these people. Within this temporal mental construct, the vast majority of the complex thought structures that *appeared* to have resulted from these alleged saviors were amassed on the basis of assumptions, presumptions, theories, beliefs, and interpretations disseminated by purported others. One of the more obvious manifestations that sprung forth as a result of these immense thought structures, that has *appeared* to mushroom over the years is our tendency to employ language to distance ourselves from what *appears* to be happening in our *seeming* reality. I'm not saying this to alarm anyone, but I see little sense in sugarcoating the situation. There's little doubt that religious thinking *appears* to be responsible for much of the conflict and division in this dreamlike construct. If one were to speak of what *appears* as the human picture, it's astounding that after so many centuries of preaching and proselytizing these beliefs that many religious constructs continue to thrive today.

Many religions preach that they can free you from bondage, when in fact anyone following a particular religious doctrine *appears* to end up living in bondage to a concept. In light of what people are, or rather, what they *appear* to be, it *seems* things can't be any other way. From a pseudo point of view within the human picture, it's difficult to see any indication that all this preaching and practicing has resulted in what was intended. In fact, it may have been the catalyst for the opposite. Surely, when speaking on the basis of the human scene, there *seems* to be more war, stealing, cheating, mental instability, fear, hate, violence, you name it, and it *appears* that there is more of it. One could say there are more people on this planet than ever before and point to this as the reason for the surge in what has been labeled behavioral deviations. The *seemingly* innate ability of humankind to rationalize and justify *appears* to have no limits. It *seems* easier to take this position than to confront what's blatantly obvious, that none of these approaches or methods have worked, and upon closer scrutiny and analysis, they are more often than not absent in the lives of those people proselytizing and promoting them. As characters in a time dream, we have no way to alter or fix a dream. In any dream one must note that the dream *appears* real from the standpoint of the vast majority of characters in said dream. The dream *seems* to go the way it goes, like the way a movie goes the way it goes. There's really nothing any individual character *appearing* in a mental movie can do about it. Even if one were to speak of a human movie, it plays out the way it plays out, each character acts its part, and has no power to do otherwise.

In terms of what *appears* as the human spectrum, there's no escaping the divisive nature of religious thinking, and the sheer volume of violence connected to religion is inescapable. Since its *apparent* infancy, religion has been used to divide and control humanity as it moves along on its purported path. Even though this *seems* to be glaringly obvious, like robots, we *seem* to continue to cling to ideologies and belief systems for comfort, reassurance, a sense of belonging, and a multitude of other reasons. There is no way to reverse this trend as the complex nature of the mechanism at work is so bent on its illusory persistence and continuity. As a character in this dreamlike mental construct one either moves along with it or one is likely to be crushed under its weight and momentum, and if this *seems* to be one's fate, then this event simply becomes assimilated into the story.

In a personal context, speculating on whether there's a god or not is a bottomless pit that ends with no real answers, so this leaves people clinging to *hope, faith, and belief.* Many *appear* to get deeply invested in a particular system of thinking and then spend their lives building upon it, protecting it, and defending it. Anything or any "body" that *seems* to come along that does not validate their pre-existing belief is viewed as a threat, and over time they become rather proficient at attacking, and or defending in order to protect their cherished position that has been built on an inherently fragile foundation. Often these complex thought structures are handed down in the family from generation to generation, so the exposure, indoctrination, and conditioning begins quite early in one's childhood. As children many of us are also sent to schools and various places of religious worship, and thus the knowledge is built on and solidified even further. Often times as children we're taught and preached to by people in positions of authority, or people we have given a certain altitude to because of what they claim to know. Inevitably, as adults the vast majority of people *seem* to continue the tradition and pass it on to their children. The process just goes on and on, and the structure grows and expands, becoming more convoluted and bloated, and more distant from its original precepts. And it's likely that even those precepts where not so original, nor does anyone really know if they had any validity. It's likely that they had no validity at all, because they all have their roots in never present thinking. They all share a commonality in that they start with an utterly false premise. How do I know? I'm not asking you to believe me, some author, just because it is stated here. I'm suggesting you find out for yourself, by yourself, that what *appears* to deny the all-ness of life, truth, reality, love — whatever word you take delight in using to speak of *all that is* — is utterly false, untrue, not real, utterly non-existent.

One who *appears* to assume one is a human being that is never actually being, in a temporal mental construct that is not being, not present is left only with *hope, faith, and belief.* It's analogous to living as a concept within the context of a relative conceptual mental construct. How is it even possible to live as: unreality, non-being, unconsciousness, non-presence? It's not! Life is now, hence, only now is alive! It is pure life itself that is consciously alive here, now.

CHAPTER XII

Final Words

If one were to speak of what *appears* as the human scene, is it possible that the hypothesized genetic mutation, what is called in this book, *the subtle shift,* is simply another genetic misstep that will realign on its own in due time? Within the context of this never present temporal construct, perhaps it was an error that nature is content to allow to run its course. Of course, from nature's view it may not be an error at all but given that human sense always starts with a divided perspective, it interprets this *seeming* genetic misstep as a mistake, or a mutation. But it would only be a misstep or mutation if one could contrast what things would be like if *the subtle shift* had never occurred. Personal sense can only speculate and theorize about what the human picture would look like today if human thinking were to fall back into its functional role as a servant. Of course, thinking would still be relegated to functioning in service to itself.

Reality is the complete absence of thinking, no uncertainty or flip-flopping, no vibration, hence no mental baggage.

When speaking on the basis of the human scene, it must be acknowledged that humans have no freedom of action or freedom of choice. In this context there is no way to escape this temporal mental construct. From a personal perspective, it *seems* that we are stuck in a temporal mental realm of opposites. Indeed, as characters *appearing* within the context of this temporal dreamlike construct, we are part and parcel to it. Even in an ordinary human dream, it's never possible for any of the characters to exit the dream. The human picture and its universe would be synonymous with dream. When it comes to any specific character, there is really no way out, so any purported person trying to get out, transcend, or get there, is getting nowhere. There is here, here (present) is now, and now is already now, already present as the present, already changeless fact. Have you ever noticed that it is never not now?

In terms of the human spectrum, the self that you know and experience as your personal self is the temporal-mind's self. The personal self or little me is not your true identity, as there is but one non-numerical reality, this means there is only infinite reality, the alone one, the true identity. Of course, a purported never present temporal-mind will reject such statements. Even though in one's daily experience of living it may *seem* that there's a passage of time, one can stop at any moment and simply notice that it is now. When speaking as to the nature of reality, word concepts can't do the job, because words are conceptual dualistic containers, and *that which is all there is of all there is,* can't be contained or captured within the framework of thought. The so-called temporal-mind is never present, and *what is,* or reality is the present, the one aware, right here, now. Thoughts about now are not now itself, again, it *seems* that as soon as we employ language we are stuck with the conceptual, and not the actual. As soon as one *seems* to speak it *appears* there is time, time that is never present, never now.

The temporal-mind spoken of in this book is vibration. The entire human picture is nothing more than a mental image, an energy pattern inclusive of what *appears* as: planets, trees, rivers, air, water, fire, space, dimension, bodies, etc. No one is really at the helm. It *seems* as though the divisive nature of thought has shifted and seeks to run the whole show. It seeks to be the game, the contents, and the audience all in one. It thinks or dreams up its own movie, and dreams as that movie. This thought movie has no foundation in reality, it can be described as never present vibration, just wisp of thought energy patterns always passing on.

What is the temporal-mind? That's the point, there is no such mind as it never really *is*. It doesn't exist at all; not even as a dream; vibration is reaction, it's not a conscious entity, it would be just reacting, passing, shifting, and changing energy. The so-called temporal-mind is not being, not present, not now, and thus not, period. It is literally a field of not. This means that all that *seems* to come forth within the context of time is not. Where did this purported interloper, this so-called time-mind come from? Time is not being, not present, not now, hence, time is not! There really is no such mind, or better said, no such mind *is. That which is not* didn't come from anywhere, as it is not present, not here, now.

Human sense can theorize ad nauseum as to the nature of thinking, but that would just be a circular exercise where thinking *appears* to be thinking about thinking. Speaking on the basis of human sense, as more lines are drawn and the divisions proliferate, it *seems* as though there is no authentic approach that can slow down the momentum of a machine that may very well mean the end of the human species — at least as we *appear* to know and define it. All of our attempts and approaches are like throwing dry wood on a roaring fire. As characters *appearing* within the context of this temporal mental construct, it *seems* we are continuously throwing thought wood on a thought fire. If the vast majority of our current problems and conflicts have their roots in a thinking mechanism that has gone off the rails, how can thinking be used as a solvent that might lead to a sustainable way of living together on this planet? Clearly this is irrational, perhaps even delusional. While this may be how the situation *appears* to our limited human senses, it also *seems* it must go on this way. It *appears* that there are no genuine solutions, as all purported solutions have their roots in human thinking, and it *seems* as though so-called human thinking is not interested in actual solutions. Speaking strictly on the basis of the human picture there is no universal solvent for humankind's dilemmas. One cannot fix that which does not exist in the first place.

In a *seeming* human context, it *appears* as though we live in an era when tremendous changes are taking place in the world. From a human standpoint, the entirety of what *appears* as our world is continuously changing, shifting, reacting energy, and therefore, change *appears* to be the only constant. In terms of the never present temporal-mind, change would be its actual activity, *appearing* as forms or energy patterns.

If one were to speak on the basis of *appearances,* it *seems* there are endless discussions and many battles debating environmental pollution, climate change, and the impact humans have on the environment, and the health of this planet in general. If one were to speak of *appearances,* our planet *seems* to have a remarkable ability to heal, transform, and evolve. That doesn't mean one should go out and intentionally cause harm. Whether one agrees or not, the human picture and its world is just a dreamlike mental construct. In a practical sense, it can be said that this planet is not in danger but, speaking in terms of what *appears* as the human species, its survival *seems* to be in serious doubt.

Frankly, knowing the nature of thought, I hesitated to write this book. I have no interest in adding to what already *appears* to be a convoluted and complex structure. Perhaps there are things that can be applied to offer temporary relief, but in all honesty, even they may end up being counterproductive, as the immensity and complexity of the global database generated through the use of thought will not yield to more thinking and figuring, it will only gain in strength and momentum. Again, I'm a realist, and in my experience, the general human belief *seems* to be that a complex problem will not yield to a simplistic solution. In fact, the so-called thinking mind or human intellect *seems* to be attracted to complexity. Essentially, this is what humankind *appears* to be up against. Thought is only interested in permanence and continuity, and thus it seeks growth and expansion, and as an offshoot of this activity, layers upon layers of complexity are inevitable, so it can't be confronted with even more complexity, it would become incorporated into an already complex construct. It *seems* only sensible that if there is any possibility at all, it has to come in the form of pure simplicity. *The true value of any principle is determined by measuring the results you get when it's applied against the simplicity of the principle itself.* That being said, we have to ask a fundamental question. If we went off on a search for a solution, a way to harness what we call human thinking, how would we know what we are looking for? It's not possible to look for something if we don't know what we're looking for in the first place. If we know what we're looking for, then we already have it. If we don't know what we're looking for, then how would we recognize it if it was to *appear?* And who is this one that is looking? It is the temporal-mind dreaming as a personal self that is looking, seeking, and searching, projecting thought into an illusory future that never arrives. It would be analogous to seeking answers to imaginary questions, trying to solve illusory problems. It *appears* to be a very strange box from which there is no escape.

If you feel so inclined, you can act as a witness to your own behavior, impulses, and responses. Pay attention to the stories you tell yourself and observe how they support the story of you and your daily experience of living. It's problematic for a never present thinking mechanism to be dramatizing, while at the same time observing. A temporal or time-mind can't co-exist with timeless awareness that is aware right here, now. It *seems* one can get very still and notice how these divisions arise in one's experience. Notice how thoughts just *appear* on their own without any personal effort, as a temporal thinking sensing mind has no personal will. Note, this is not a method, a ticket to freedom, or a get out of jail free card. Sometimes observing instead of dramatizing leads to a sense of uneasiness and a feeling of instability, it's as though the machine senses the danger to its own illusory survival and continuity. Again, this is not a remedy or panacea to the dilemmas we *appear* to be currently facing, but one never knows, perhaps something comes out of it. Maybe what we call nature is already in the process of applying a remedy while we *appear* to be busy searching for one. Perhaps it's automatic and built into nature inherently. Maybe we'll just be recycled like everything else, and what we see as a problem is not really a problem at all, but instead, relatively normal. If what *appears* as a never present human picture is vibration, energy randomly reacting, shifting, and realigning, moving along in its own dreamlike fashion with no one at the helm, then it's likely that there is no solution. Frankly, any potential solution that arises from thinking cannot be a solution. The truth is that it *seems* that there is nothing we can do. I know that sounds cynical but let me add to this by saying that perhaps there isn't anything we need to do. It's conceivable that whatever we think or believe needs to be done has no meaning at all. Starting with the assumption that one is a personal separate little self with a separate mind, functioning in a physical reality that is always moving along with the passage of time, it *appears* there is a lot one can do or needs to do. Starting from now, indeed, as now, what's done is done and never has it been undone, never will it be undone, for it is eternally now. Life simply *is,* and starting with this premise, there is nothing that needs to be handled, fixed, resolved, healed, helped, or treated. The inherent intelligence that *appears* to be underly or be superimposed upon by a myriad of ideas, concepts, and mental constructs is forever changeless. Now, the present, doesn't come and go, or ebb and flow. Reality, *what is,* is not a special kind of energy, a vibration with a frequency. Pure awareness simply, effortlessly is *all that is.*

All the learned knowledge accumulated over time within the context of the *seeming* human spectrum is dwarfed by the inherent intelligence that is right here, now. It's not about words or concepts, for all of our ideas, concepts, and theories will never help us to capture or comprehend the true magnitude of this intelligence. We can come up with an assortment of names, definitions, theories, and interpretations, and this is where we hit a dead end. The moment we name and define something we have put it into a container, and when speaking about or attempting to point to this intelligence, clearly there is no way to define, contain, point to, or grasp exactly what it is, nor can we personally experience it, much less give expression to it. The immensity and the immediacy of this intelligence does not allow for any movement toward or away from its own totality. Language fails to have any practical usefulness in regard to this kind of discussion. Even the word god is a container, and then one inevitably gets so-called people arguing over their particular concept and interpretation of what they *think and believe* god is, and what they *think and believe* god is not. Neither side really knows — as a finite mind cannot comprehend the infinite — so they are each stuck living in bondage to a concept, and reliant on *hope, faith, and belief* as they move along on their path to oblivion.

Where exactly does this leave us? It leaves us right where we're supposed to be. How do I know that? Because that's where we are and that's the way it *seems* to be. Speaking on the basis of what *appears* as the human picture, we *seem* to be characters in a temporal dreamlike mental construct. The entirety of our world and our experience of it is all mental, nothing but thought. If one is able to accept the ephemeral nature of one's *seeming* status, a fresh perspective *appears* to arise organically, allowing one to understand human experience in a broader and more realistic context. Perhaps the sub-title of this book is not so ominous after all, but it certainly can *seem* to be if one clings to a manner of thinking that *seems* to suggest one's identity is merely a separate little self, stuck within never present time, a temporary personality that is destined to meet an inevitable end.

This purported temporal dreamlike mental construct *appears* to be the only *seeming* reality we can deal with, and we must deal with it on its terms because we are inseparable from it. Fight it, argue with it, resist it, do what you will, but know that you will lose. Some have suggested a giving up, or what is sometimes described as a total surrender to *what is,* but even that approach is questionable, as one would be giving up in order to get something or somewhere, and that's not really an absolute surrender in an unconditional sense. One cannot make an unconditional surrender within the context of a conditional construct, and the idea that a complete surrender, a giving up, would result in some kind of change or shift that would free us on the instant is simply an illusory carrot aimed at inciting even more thinking. After all, given what we *appear* to be within the context of this temporal mental construct, just who is it that would be surrendering? And just who is it that needs to be freed? This so-called approach or method, if you will, *appears* to be another game the temporal-mind *seems* to be playing with itself to ensure the continuity of its pretense.

The following statement below is made with a certainty that can't be conveyed to another, and no one is suggesting that one should blindly believe it at face value. One is free to do as one pleases. Truth is true only to truth itself, because it truly *is,* it is already fact. If what is declared is truth, whether or not it is accepted or rejected is meaningless, as truth is not the result of something nor does truth result in something. Truth is not something one seeks or discovers. Truth is now, so only now, the present is true. Now, reality, truth, are synonymous terms.

Infinite intelligence is pure now, it is existence itself, nothing caused it, nothing can threaten or challenge it. It has no beginning and no end, no inside, no outside, no center, no circumference, and any effort put forth to touch or grasp it via a purported human mind is futile. If we are but characters having the experience of a beautiful dream, then why would anyone want to be awakened? In truth, who is there in a dream that can be awakened? Clearly, it *seems* that for some the so-called dream has become a nightmare and so I publish this work. Perhaps something comes out of it. Perhaps someone is touched in a way that triggers a different kind of shift, where the *seeming* pseudo identity, the *apparent* personal self is annihilated as though having never existed, because in reality it never did, does, or will. Life is now! Now is not a time, not in time, nor a product of time. There is no way to now, nor is there anything in the way of now. *Only now is!* This means one is now or one does not exist.

Author's Message

My wish for you is that in an instant without warning — something akin to getting struck by lightning — that the ineffable aliveness that is present right here, now, as the present, is invisibly palpable in your daily experience of living.

About the Author

Paul Richard Hillson writes on various subjects but his primary focus is on the nature of reality. He has studied this field for over 45 years. In fact, his interest in this subject has been a constant companion since his adolescence. The number of books read, and the various philosophies investigated are too numerous to list here.

Although Mr. Hillson seems to have had many exploits and delved into numerous approaches and methods, he is adamant in stating that this book did not come about as a result of that journey, but seemingly in spite of it. Mr. Hillson feels that the experiences he has had, what he's accomplished and where his inquiries have taken him may potentially serve to create a bias toward a particular perspective, so little of that is mentioned here.

Mr. Hillson worked as a designer for over 25 years - this included print and web design, copywriting, editing, and marketing. He currently resides in California, where he continues to write and enjoys living a life of simplicity.

Other Books by this Author

TRUMP And The Hidden Mechanism That Controls Humanity

www.ingramcontent.com/pod-product-compliance
Lightning Source LLC
Chambersburg PA
CBHW020904180526
45163CB00007B/2622